ESTIMATE, GUESS, OR PROMISE?

DOUGLAS R. ALLEN

authorHOUSE®

AuthorHouse™
1663 Liberty Drive
Bloomington, IN 47403
www.authorhouse.com
Phone: 833-262-8899

Published by AuthorHouse 08/17/2020

ISBN: 978-1-7283-6801-6 (sc)
ISBN: 978-1-7283-6799-6 (hc)
ISBN: 978-1-7283-6800-9 (e)

Library of Congress Control Number: 2020913496

Print information available on the last page.

This book is printed on acid-free paper.

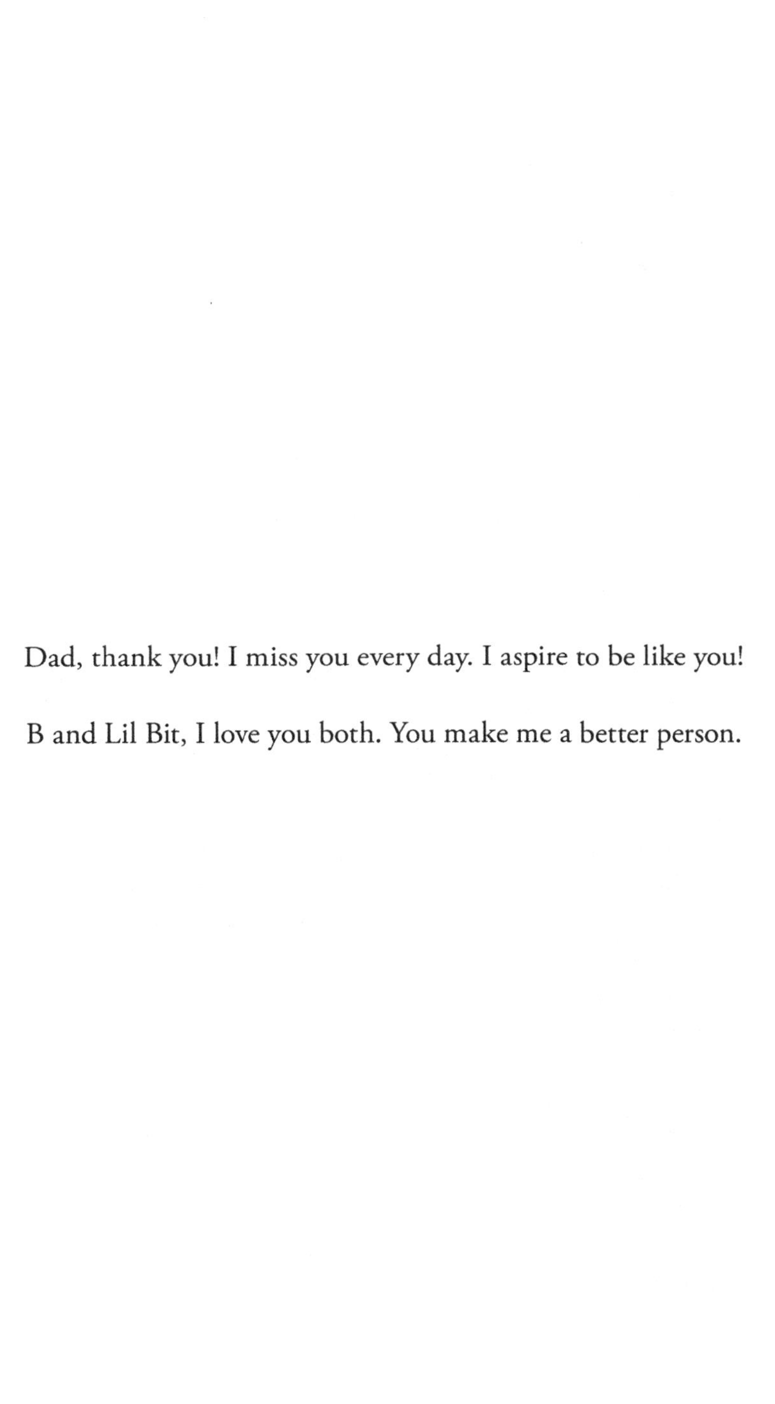

Dad, thank you! I miss you every day. I aspire to be like you!

B and Lil Bit, I love you both. You make me a better person.

INTRODUCTION

A few years ago, I decided to take my wife and our four children on a fourteen-hour journey to visit my family in Virginia. Not to date myself, but this trip took place before the era of mobile devices, meaning my wife and I did the entertaining by settling arguments and making idle threats hoping to scare the children into behaving. While driving and considering the legalities of placing the kids in the truck for the remainder of the trip, I noticed the welcoming sign of Cracker Barrell. The objective, feed them, and hope they sleep for the rest of the journey.

After a hearty meal and a lengthy debate with my wife on gift store purchases, I approached the cash register to pay for the meal and my new souvenirs. Standing there, waiting for my receipt led to an inspirational moment of brilliance, which I knew might keep the car silent for a few hours. Right behind the register on the shelf set a jar full of the brightest looking candy I had ever seen. I had a great idea!

The kids suddenly became well behaved watching my every move to see what or where I put that candy. If you are a parent, you know what I had, currency! As a parent, leverage is the key to controlling small children, and I had it. I handed the jar to my youngest boy and told him to hold it for a minute and then pass it on to the other children. The excitement overwhelmed them, but I knew I had to hold out to keep them behaving. The pressure mounted!

Just when I started to see their interest sway, I told them we would play a game and whoever won would get all the candy in the jar. Of course, I am making the rules up as I go, but it worked. Each kid would receive ten seconds to touch the glass container, shake it, do whatever they needed to do to determine how many pieces of candy it contained. The kid that guessed closest to the number of candies won the entire jar. My wife even wanted to participate! Feeling like Dad of the year, I said, "let the games begin."

With four kids ranging from age five to ten, and my wife participating, I knew this game would be a game to remember. The oldest took the jar and held it for a few moments as if she was performing a weight calculation to determine the amount of candy. The second oldest studied the jar, muttering under his breath what seemed to be calculations. The two youngest looked at the jar and then looked at everyone else as if to say, "Please help me. I just learned how to count."

Just as I started to wonder how I would come up with an actual number, my wife already ahead of me had opened the jar and started counting the candies. She came up with the total as the anticipation amongst the children increased.

The moment of truth arrived. The children eagerly awaited the results as she compared the guesses against the actuals. By the look on her face, I quickly realized we did not have any child prodigies! The range of estimates varied significantly with none of them even close to the actual number.

Years later, this seemingly silly story came back to me as I worked with professional estimators in construction companies. My consulting career in cost accounting had somehow led me into managing a construction company, working with estimators to push competitive bids out just about every other day. I could not believe what I observed, these individuals or estimators each worked independently to arrive at the appropriate number for the bid and then turned it over to someone else to compile it all into one number. As a novice in estimating, I remained silent, but in my head, I thought this is too risky! Just as I expected, in the final moments before they had to send off the bid, the room went crazy as the different estimators discussed material variances between the various inputs from subcontractors and internal self-performance calculations. Ironically, without hesitation, one individual made the call submitting the bid at the last second.

How did all of these professionals review the same documents, speak with the same vendors/subcontractors, follow the same specifications (including a standard wage rate), yet all of them ended with different conclusions? At that moment, the silly little story with my kids started to make sense. In the car, my kids all incorporated some method to estimate the number of candies in that jar. Each of them felt good about their process, yet none of them were accurate. Ironically, that sums up the estimating approach of many estimators in the construction industry.

Webster defines the word estimate as "to give or form a general idea about the value, size, or cost of something." Ironically, Webster defines guess as "to form an opinion or give an answer about something when you do not know much or anything about it." Unfortunately, sometimes these two definitions merge into one, and construction companies roll the dice on a bid hoping for the best outcome. The point is not to question the expertise of a professional estimator, but rather to consider the entire process by which companies arrive at a final number to submit. Using the word estimate conveys a certain level of knowledge or expertise, but based on my observations many times, the end product could fall under either of these definitions.

The approaches my children took when they estimated the number of candies in the jar remind me of behavior I observe from many professional estimators. My eldest child figured her guess would be superior to the other children strictly because she had been around longer and a few years of education under her belt. Some estimators, like my eldest, make assumptions purely based on tenure, arrogantly placing numbers in the bid with little to no substantiation. If they are wrong and their number is too high, resulting in the award going to someone else they say, "that other contractor missed something, or the other contractor is buying jobs." On the other hand, if they come in low, resulting in a win but a potential financial loss, they start talking about all their years of experience and how this must be a mistake. I lovingly refer to this estimator as the "know it all that knows very little."

My second to oldest handled the jar very cautiously appearing to perform a calculation of sorts with hopes of maybe psyching out the other kids. His response reminds me of those estimators that know how to play the part, talking as if they understand the job, running various calculations and making it appear as if the numbers they provide a result from tried and true methods. This type of estimator depends entirely on their individual experience, ignoring other inputs, and refusing to accept information from third parties, subcontractors, or suppliers. Once they decide what the cost for the job are they are done and will listen to no outside sources. The estimator in this role never seeks advice, rarely asks for assistance and, most important, does not like being questioned. He/she might say things like, "We can change it, but if it is wrong, I want everyone to know it was not me." By far, this individual is the most frustrating person to deal with because they are a gift to the world of estimating. I refer to this individual as the "lead actor in a single star production."

My two youngest kids, unfortunately, represent many estimators out there bidding jobs right now, great people with lots of effort but no competency. My littlest ones had no idea even where to begin figuring out how many candies were in the jar, so they both looked to their siblings for help. Many estimators do this on every bid, accept numbers directly from subcontractors without ever even challenging them. They take on the attitude, "They know everything; I will just take their number and mark it up." These people are always the type asking for contingencies or looking for ways to "make it up" on other line items. This mentality permeates the industry; many contractors estimate based strictly on a statistical premise that if they bid enough, despite the inaccuracies, they will win enough to maintain operations. You will hear these folks say things like, "I think everyone is going to bid around this number." They are continually trying to figure out who other people are using or are concerned about who might be bidding. In many cases, these folks spend more time trying to figure out why not to go after a project than how to strategically approach the project. I refer to these folks as the "if you can't beat them, join them" group.

The final person in my story is my wife. Initially, she intended on guessing until she realized that without an accurate count, there is no way I would have known if any of the kids guessed accurately. She opened the jar and counted each piece of candy, providing the most accurate estimate possible. The estimator like my wife keeps searching, asking lots of questions, documenting everything, and verifying every number through multiple resources. They compare all the input, develop a relevant range, and then perform additional calculations, ensuring accuracy. He/she listens to all input and is more concerned with accurate information than their reputation or their ego. All they care about is finding out what the real cost is for completing a specific task or line item. They do whatever it takes to be accurate, just like my wife taking the top off of the jar and counting the candy. Guess what I call this individual an "estimator."

Ironically, the industry is full of individuals who possess the title of estimator but do nothing of the sort. The purpose of this book is to assist those who seek to estimate, not guess, not pretend to know everything, not be so proud as to ignore inputs, and not throw together numbers with the hope of winning based on some preconceived notion that because you have done this for a long time you deserve an award. If your objective is to create accurate estimates, then read and apply the principles presented in this book and success is in your future.

There is an excellent book out there by Michael E. Gerber called The E-Myth Contractor, in his book, he speaks to what I am talking about in this section. As a contractor, one must provide much more than just an estimate, "…we no longer provide estimates. Now we define exactly what it will take to do the job, how much it will cost, and how long it will take, and then we give you our written promise that what we have agreed upon is exactly what we will deliver" (Gerber 2003, 159). As a former contractor and now a consultant for sureties, it amazes me the risk contractors assume to win a project. Unlike other industries which continuously adjust to risk or work to mitigate it, contractors seem just to push forward, hoping for the best outcomes.

Let me illustrate by talking about an industry we all know is in a constant state of learning - the insurance industry. In the late 1760s, the insurance industry began selling life insurance policies to cover widows and families of ministers. Today, insurance companies include everything from body parts of celebrities to just about any asset imaginable. The insurance industry successfully does this because of their commitment to understanding risk and preparing estimates to address the probability of loss accurately. The industry possesses a group of individuals referred to as actuaries who do nothing but study risk to understand how to mitigate the risk and ultimately gain a return. On the other hand, the construction industry risk continually increases, yet we still prepare estimates and act as if it is the 1760s! The estimator, in my opinion, must become like the actuary, examining the project, understanding the risk, and advising management on how to mitigate risk and make a suitable return.

The bidding process is no longer an abstract guessing process, but rather a scientific, calculated approach to the performance of a task. It is a guarantee from you, the estimator, that the amount of candy in the jar is accurate because you counted it! You applied the most tedious processes to ensure accuracy, actual cost, and ultimately a promise that, if executed correctly, a return is waiting. This type of bidding creates confidence, wins projects, maintains contractors as a going concern, generates a higher kill rate, delivers consistent margins, provides steady workflow, and creates a reputation of excellence in the industry. If that is not enough, consider the value of your position within an organization when you deliver!

One final note, most of the ideas in this book do not align with the industry's current approach to estimating. Ironically, estimators typically agree with the theory, but as soon as they begin bidding, they revert to utilizing secret spreadsheets and do not consider the time element of a project relevant. When you start using the methods, I propose in this book, be prepared to hear negative feedback from the "seasoned" professionals. If you think like me, you will appreciate one of my core values that keeps me moving forward in an industry that seems not to want to change: "Never act on the premise that it has always been done this way!" I challenge you to approach everything in this manner, questioning every aspect of what you do so that you do not become one of those people who do things because it has always been done a particular way.

CHAPTER 1

Go or No Go?

What I am about to discuss seems logical. It is so obvious you might even wonder why I make it my first point. Unfortunately, as humans, we ignore undeniable facts to pursue the path of least resistance. Next time when you are sitting on the plane waiting for everyone to board observe the pilot walking around the aircraft examining different mechanical items. The captain of the flight does this inspection with the hopes of identifying potential issues before take-off. Pilots perform the checks based on written procedures outlining what to inspect and the parameters. The written process mitigates the risk of accidents by monitoring the human flaw of skipping or forgetting to look at a component of the plane. In the bidding process, the contractor must develop a written procedure to prevent another type of disaster of the financial kind.

While writing this book, a friend of mine asked me an interesting question, "If you could only give one bit of advice to a contractor about his bidding process, what would you say." Without hesitation, I told him the first thing any contractor should do is determine if he/she should even pursue the project. Sounds simple, yet this is the number one missing question that leads to the demise of many construction companies. The go/no go decision is the crucial first step in delivering the promise to the customer.

In the surety business, personal character is an integral part of the evaluation process to decide whether to issue a bond on a project. I do a presentation to underwriters, where I give them clues on identifying a contractor with issues related to character. The first point of the presentation addresses contractors who exist with no game plan, no strategy, just going after whatever comes along. When you observe a contractor operating in the above manner, it is only a matter of time until they pursue a project that ends in disaster. After all, remember in contracting you are always one job away from shutting your doors.

In Government contracting, there is a program referred to as 8(a). If you don't know what it is, let me explain. It is a program the Government sponsors to assist minority contractors with developing a sustainable business model. It enables contracting officers to award contracts without requiring the company to compete. Now that you understand, let me tell you a story about an 8(a).

Contractors do not relate to the idea of No-Go! It is not in their nature, as a matter of fact, it is directly opposed to their chemical makeup. Turning down a project is just not in their DNA. In the latter part of 2010, I had a good friend working in contracting as an 8(a). Billy Bob (made up name) owned a small construction firm with an excellent reputation. He performed, jobs completed on time, quality work, and with great pricing. Honestly, I envied his position and his capabilities. On top of all it, he had the 8(a) status meaning contractors could go directly to him for work.

Every few weeks, Billy Bob and I met for lunch. He knew the future looked bright, and he walked me through about ten different success scenarios. Within fifteen minutes, Billy Bod laid out twenty different ideas related to building his company. Finally, he stopped and said, "I need your help, Douglas. What do you think about what I told you?" I said, I think it sounds exciting, but I am a little concerned about your lack of strategy as far as it relates to acquisitions. Make sure you develop a detailed acquisition strategy before pursuing projects. He heard the words, but no processing!

About three months later, I heard from the SBA Billy Bob convinced the Corps of Engineers to sole source a massive project to his company under the 8(a) program. I remember thinking that he did not possess the capacity to bond a project of this size, but I figured he must have figured it out or he would not have pushed to get them to sole-source the project. I sent him an email and congratulated him on his success! I told the SBA agent about my talk with Billy Bob, and we both hoped he had listened.

Sunday night at a little after ten, I noticed someone had been blowing my phone up. Six voice messages all from the same phone number. Just as I started listening to the voice messages, the other line showed a phone call. "Hello, Douglas? Billy Bob here. I need to talk to you."

After the niceties, his voice quivered as he explained his situation. Despite my advice, he accepted all kinds of work, not in line with his capabilities and ended up buying his company out of projects, which hurt him financially. He told no one, thinking that he would win this project with the Corps of Engineers and then do what I told him – develop an acquisition strategy. It turns out, he lost his bonding capacity. One of his projects was in claims with the surety taking over the project.

Finally, he got up the courage to ask if I would assist him with bonding the job. Even if I had been able to help, the Corp would never go for another contracting bonding the job. Also, I knew my agent would not consider bonding a contractor in this situation. The next morning, from what I understand, he went to the Corps told them what happened, and they gave the work to another contractor. After another six weeks, Billy Bob shut his company down and went to work for a competitor.

Sad story, you know what it is even more disturbing? I could tell you about fifty more of these same stories. Contractors with so much potential, yet they pursued everything coming at them until they lost control. You probably read the story and thought, "What an idiot!", why did he go after work without the proper bonding. Well, why do contractors chase work without the appropriate resources, or the experience and knowledge to perform, or the right team to complete a job.

Outcomes such as this are avoidable. Billy Bob should still be in business, working a strategy for acquisitions, growing organically, and building a stable company. Instead, his business ended, reputation done, and he now works to pay the surety back for finishing the project. Lesson number one, contractors must learn to say, No! No, that job does not align with our strategic plan! No, that work is not something we know how to do! No, No, No! Say no to the wrong jobs so that you can say yes to the right ones!

Just like the pilot doing the inspection, let me walk you through what you need to do to avoid a crash. Let's walk through what it takes to make sure the plane is flight worthy, or in the case of contracting the project is worth pursuing. It is as simple as understanding go/no go!

Know thyself....

When I lived in Nevada, I fell in love with cycling. Yea, I know, riding outside on skinny wheels wearing nothing but your underwear...really cool! I quickly fell in love with riding, and because of my obsessive tendencies, I got involved with criterium racing. If you don't know what a criterium race is, I would recommend you Google it, but for a quick description, it is like NASCAR for cycling. Each track contains four corners with the entire course under a mile. About fifty to one-hundred people line up and race around the road for about forty-five minutes. The first person to cross the finish line at the end of the time wins the race. It sounds simple, but I challenge you to go out tomorrow in your underwear and ride your bike for an hour at a pace of around twenty to thirty miles per hour and then see if you think it is easy!

One lesson I learned quickly in the sport is strength does not equate to victory. The strongest rider, or in most cases, the person out front during the race seldom crosses the line first to claim victory. The winners standing on the podium usually include the guy/gal that exerts a lot of energy and gets out in a breakaway (far away from the peloton) or the guy/gal sitting in the back until the last second before putting forth significant effort on the final lap, referred to as sprinting.

The first race, well, I embarrassed myself. The whistle blew, and I pedaled as hard as I possibly could for about two minutes and then went into a mild form of cardiac arrest (it felt like it). As my pedals slowed, the peloton of one-hundred bikes breezed past me. When the race ended, my coach looked at me and said, "The key to racing is understanding what you can do well, and what you suck at doing. Be honest with yourself." At that moment, I listened, but it took a while for me to understand what he meant.

Competitive racing requires an understanding of both strengths and weaknesses. Knowing your limits enables you to work both ends of the spectrum to your advantage. As a sprinter, my capabilities include quick explosions of high wattage, not sustained long pulls. When I race, my objective is to use as little energy as possible, conserving it for those small moments during the race when I need to put out some serious power. If things go as planned, at the end of the race, I am full of energy and ready to lay down some serious wattage to cruise across the finish line.

On the other hand, there is my skinny buddy that enjoys maintaining a steady wattage, or power, for an extended period. If he even attempts to sprint, his race ends immediately. Two different strengths, but each of us are in the same race. In cycling, you must understand and leverage your strengths. If you don't, you end up frustrated with a group finish, or even worse get dropped!

You might be thinking, what does that have to do with anything! Well, I am glad you asked. As a contractor, it is imperative to understand both your strengths and weaknesses. Bidding, like racing, is competitive. To be competitive, you need to know your strengths and leverage those in the bidding process. In the middle of the race, it seems cool to lead the pack, but in reality, it is just for show. If you attended a race, and you knew me, you would be thinking, "He'd better get back, or he will run out of energy." In bidding, I watch companies pursue projects that are way out of line with their capabilities, and I think, "What are they doing, that is not their strength." Other people, bystanders, may say, "Wow, what a good contractor." When you pursue a project without any consideration of your strengths and weaknesses, you are cruising up front wasting good energy/resources for nothing.

For twenty-plus years, I have worked with and observed all types of businesses in just about every industry. No matter the industry, it never ceases to amaze me the lack of congruency within an organization as it relates to their strengths and weaknesses. In 2010, I did some consulting work for a relatively large company. I met with the senior leadership and discussed what they envisioned for the future, goals, tactics, and general strategy. They also shared what they each felt set the company apart, or their strengths, and also what needed improvement, or weaknesses. They told me all of this, yet the reason for my engagement is because the company was in disarray. I asked if it would be okay to speak with the middle managers and ask the same questions.

Armed with a notepad, I began meeting with the various department managers. I asked each manager to share their opinion of the vision and what they considered strengths/weaknesses. My only regret, I did not get permission to publicize the results because what they shared is excellent! Based on the performance of the company, I knew none of them would share the same vision, but I was shocked to find that the strengths/weaknesses were utterly different from what upper management shared. Here is a large company, pursuing all kinds of projects with no understanding of who they were! Surprisingly enough, in a matter of two years, the company shut its doors because it bid a few jobs, not in line with its capabilities - surprise!

When bidding, the entire team needs to understand precisely what defines the strengths and weaknesses of your organization. Right now, if you ask your bidding team the strengths and weaknesses of your organization and they don't know or answer differently than what you think - you got a problem. There is an excellent chance you might pursue a project, not in line with your capabilities. Bottom line, companies waste valuable resources pursuing work that is going to do nothing but erode margins and eat away at the core of your business.

The obvious next question is, how does one go about identifying strengths and weaknesses within an organization. I am so glad you asked! Below are two easy methods:

1. SWOT Analysis (Strengths, Weaknesses, Opportunities, and threats)
2. Project Autopsy.

The SWOT analysis is not a new concept; however, most contractors and businesses, for that matter, never utilize it fully. Unfortunately, companies and more specifically, contractors perform the analysis but never apply what they learn. Remember, we are using this to decide whether to pursue a project or leave it alone. In many cases, you will realize the strengths identified on a project also represent your strengths as an organization.

Strengths and weaknesses relate to internal matters, whereas opportunities and threats deal with external forces. It is essential to keep these separated as you perform the analysis, for reasons I will explain after I go through the four categories.

Your team should begin the SWOT analysis by collectively brainstorming recent projects to identify internal strengths. What components of the last project under your direct control really stand out and make your company look great? A subcontractor is not an inner strength, but your ability to manage multiple subcontractors might be something to highlight. Make sure you involve every single team member (all four of them—I will explain later) in the discussion and document the entire evaluation process. Do not argue or let anyone question the comments of the team. Write down everything everyone says without modifying or condensing. Below are some examples of internal strengths your team might mention:

Scheduling	Project Management (Name of Person)
Self-Performance	Proposal
Customer Relations	Budgeting

The next topic, weaknesses, is not a favorite of anyone, especially those working on the projects discussed. Make sure you keep the dialogue open during this conversation by not attacking people, but rather problems. Start the conversation with your team by explaining that every project inherently comes with weaknesses. The objective is not to raise old issues, but to identify areas needing attention. For example, on a project my company performed OSHA showed up and dinged us for some minor infractions. To me, this illustrated a hole in our safety plan, a weakness needing attention. It is imperative to approach this honestly, not dismissing minor details because sometimes the small things give insight into your next challenge on the job!

Opportunities and threats focus on external elements beyond the control of the contractor. Sometimes people skip over these two because they consider this out of their control. The objective is to evaluate these external factors and determine how to leverage your strengths to overcome threats and seize opportunities to address weaknesses.

Opportunities enable the contractor to identify potential tools to overcome weaknesses on a specific project or for the entire company. For example, on your last project, the team struggled with maintaining a good schedule, which frustrated the owner. In the last few weeks, there is a new vendor in town offering to manage project scheduling and produce the necessary reports to satisfy the owner. Guess what? You just realized an opportunity to mitigate a weakness. Or what if new technology is available to assist with the frustrations of daily reporting. The point...look at what's available and see how it ties in with your strengths and weaknesses.

	Strengths		**Weaknesses**
S1	Project Management	W1	No self-performance
S2	Financial Resources/Bonding	W2	No equipment (rental only)
S3	Reputation w/owner	W3	Paperwork/accounting/labor reports
S4	In-house scheduling	W4	Geographic location
	Opportunities		**Threats**
O1	4 man drywall crew - no finances	T1	Labor Pool
O2	Competitor went out of business	T2	Project Delays
O3	Joint Venture w/ large contractor	T3	Scope Creep
O4	Multiple projects / same owner	T4	Competitors (bid low - buy jobs)

The last part of the analysis addresses external threats. Again, this is dealing with issues outside of the company, matters outside of the control of the organization/company. Threats may come in the form of a weak subcontractor based, difficult owners, or any other risk that impacted recent projects. Years ago, I owned my first company right around the same time Hurricane Ivan devastated the panhandle. The lack of available labor after the storm posed a threat to my business. Take the time to think this through carefully. Many companies skip right over this section and miss out on identifying significant risks. If you figure out the risk outside of your control, it might save your next project or potentially your company.

After you compile all this data, remove the duplicates, and prepare a list for each of the four categories. I would recommend using a large whiteboard so everyone can see what each category contains. First, go through the strengths and talk about how they align with the project you are considering, then do the same thing with the weaknesses. Talk through these points, see how they play into the new project. Do they align so nicely with the proposed project it is an absolute go, or do the weaknesses pose a significant risk based on the demands of the project? Do the same thing with opportunities and threats and talk about how this impacts the project you currently are pursuing. The SWOT exercise engages each team member requiring them to consider whether this the right project to go after, or one to avoid.

Below is an illustration of a SWOT analysis for a typical construction company. Observe how the various categories work together to assist with making the right decision related to pursuing a project. After the illustration, I will walk you through the process.

Please do not laugh at my pretend company SWOT analysis. Just pay attention to how this works. The objective here is to see if you can leverage your strengths and opportunities to mitigate the risk and threats. When you are doing this on a whiteboard, write the numbers beside each one if it helps. Consider W1, no self-performance. On the last job, the lack of self-performance impacted performance, so look at what you might use to overcome the weakness. The first strength, S1 - Project Management, indicates maybe instead of struggling with laborers, the company should manage subcontractors. If you focused on this strength, it might negate the weakness of self-performance. How about leveraging O1, four-person drywall crew no finances. What if you applied another strength, S2, and funded the four-person drywall crew to work directly for your company. One more, what about weakness W3, paperwork, accounting, labor reporting. How about leverage O2 and see if any skilled office personnel might need a job.

If you do this, and you do not see a way to leverage your strengths and opportunities, then you need to think about whether the project you are looking at is worth the risk.

Of all the tools incorporated over the years to assist in understanding a project, the autopsy is my personal favorite. Science my brain do not mix well. If I had no choice but to enter the medical profession, I would become a coroner. Yea, I know what you are thinking - weird. Maybe so, but think about it, I would never hear any complaining, I could run whatever test necessary to determine the cause of death, and I am pretty sure the malpractice insurance is reasonable. After each project, you need to perform an autopsy to determine the cause of death.

Before you do the autopsy, you need to understand the rules:

1. The project is dead (over), meaning there are only facts to discuss, no excuses or rationalization.
2. The purpose of the autopsy is to determine the cause of death. What happened in the life of the project that contributed to its eventual demise, good/bad?
3. Did the project die due to natural causes? Did everything go as planned? If not, what happened. If so, why did it happen?
4. Family members should not perform autopsies. Do not have this done by the team running the project. They are too emotionally connected and might not do such a great examination.
5. The autopsy is not for a murder victim. It is not a witch hunt. If someone on the project team performed poorly, you need to address them separately.

This exercise requires the team to dissect similar past projects to determine if moving forward on this potential project is a worthwhile endeavor. The past is the absolute best (the only) predictor of the future as it relates to project success.

Completing the autopsy requires the team to dissect past projects to determine if moving forward on the potential new project is a worthwhile endeavor. Unfortunately, the past is the absolute best (the only) predictor of the future as it relates to project success. Listen carefully to what your team tells you; go through the project with a skeptical attitude.

My company performed this autopsy once in consideration of an asphalt project for the Corps of Engineers. The team ran through the most recent projects and agreed our competencies did not line up with moving forward. I, being so smart, overrode the team, ignored their concerns, and talked them into pursuing the project. Almost one year later, I regretted persuading the team to pursue the asphalt project as I negotiated with a subcontractor to redo our asphalt work. None of this works, if you do not listen!

Two final points on the autopsy: First, projects, like people, die of natural causes. When a project ends just because it went great does not mean you cannot learn something from it to improve upon. Second, do not forget about the SWOT analysis and the strengths and opportunities, which might open up some avenues for going after a project despite the analysis from the autopsy.

These two methods should become ongoing processes occurring numerous times throughout the business cycle. As an owner/manager, searching for trends should become routine. If a company keeps making the same mistakes, generates low margins, or continually struggles with budget or time, then further evaluation is needed. Conducting meetings on projects not only assist with choosing the right project to pursue but also supports business growth. Verne Harsh in his book *Rockefeller Habits*, states, "Priorities, data, and rhythm are the key tools for handling the barriers that come with growth and keeping the company aligned" (Harnish 2002, p. 81).

One final note as it relates to understanding your company. Make sure everyone in your company knows why you exist, and what is most important to you as the owner (core values). Without these basic concepts in place, people within your organization make decisions with no foundation or without consideration for what is important to you, the owner. Every individual within your organization should be intimately familiar with your overall approach to business. When your team evaluates a project, the first filter should ask, "does this line up with our mission and core values."

Sometimes we do not appreciate concepts such as core values and mission statements until we observe them in practice. I remember the first time I watched these concepts applied in a business setting, and it made an impression on me to this day. From the minute I joined the firm, the organization did a great job of putting the mission statement and core values front and center. We review them before every meeting, and they were plastered all over the office. They did a superb job of making sure everyone understood the purpose of the firm.

After about a year of service, I joined a committee that evaluated new opportunities for the firm. During one meeting, a marketing representative introduced a potential audit with a reasonably large adult entertainment company. Within a second of the marketing guy finishing explaining the opportunity, the manager in charge stood up and said, "No matter how great this might sound, or the potential money, we cannot do it because it is in direct opposition of our core values, next!" When your people understand the core values and the mission, they will protect the company from the wrong project.

By now, two things happened with the team. First, every team member is fully engaged, thinking, planning, and evaluating the potential project. Second, each of them also is fully aware of the companies strengths, weaknesses, areas for improvement, and how all of it needs to come together to make this bid a success. It is a little thing called strategy forming in the heads of your team.

Know thy customer…

Are you ready for a bombshell? Here it comes...in my younger years, I preached! Yep, I studied Theology and intended to change the world as a Baptist Minister. I still want to change the world, but I plan on doing it a little differently! When you are studying for the ministry, public speaking becomes job number one. My college did a great job of setting up opportunities to speak, I even spoke at commencement, which hosted about 8,000 people. For nearly two years, I spoke publicly three times a week at a minimum.

I spoke everywhere, nursing homes, prisons, small churches, large churches, the fair, camp meetings, graduations, kid meetings, adult meetings, business gatherings, and for sports teams. Occasionally, a semi-retirement home requested speakers, which came with a small stipend. The letter arrived in my school box, giving me all the details about the speaking engagement. It included the number of people, the median age, and the mental condition of the crowd. According to the document, the fifty people attending operated on full mental capacity.

Total excitement for the opportunity led me to hours of preparation. I wrote the most eloquent message on God's never-ending forgiveness. I planned a dramatic ending with an altar call allowing those present to do business with God! Unfortunately, the crowd of forty-eight souls lacked the mental and physical capability to even stay awake for my introduction. After I finished the final prayer, no one in the room opened their eyes; they either fell asleep or worse!

The front desk clerk apologized and told me they mixed up the information and sent me the wrong card. Oh well, I left there feeling humbled and a little embarrassed. One thing I learned for sure is the importance of knowing and tailoring your message to the audience. Good speakers understand the audience and prepare to speak directly to them. Poor speakers lose their audience quickly because they did not take the time to know them.

Pretty sure, you picked up the point in my illustration. In bidding, you must know your customer. As a Government contractor, we bid to the Navy, Air Force, Corps of Engineers, Department of Forestry, the FAA, and a bunch of other agencies. Guess what? Every single one of them is entirely different from the others. Each represents a unique audience. How you deal with one of them is not anything like dealing with the other one.

Before you bid, or start developing relationships for bidding, make sure you do your homework on your potential customer. Find out everything you can about the company/agency. Do they pay on time? Are they difficult? I could tell you story after story about companies that win and then find out the payment terms are sixty to ninety days. Know your customer!

Understanding your audience might be the best advice of this entire book. In my current job, I deal with companies every day that jumped into a relationship with an owner, a large contractor, or a government entity without understanding what is required. Several years ago, I consulted with a firm that had recently acquired an IDIQ with a boss contractor at a military site. They retained me to review the specifications and assist them with preparing all of the various packages required upon award. In our first meeting, I listened to them as they explained how perfectly the work aligned with their core competencies and how they strategically bid based on self-performance capabilities. After about thirty minutes of listening to their expose on victory, he pushed the file across the desk, which contained the RFP. I opened the cover and said, "I did not know you were a union contractor." Again, deer in the headlights! They did not understand the client or the location and let me assure you they paid dearly.

Entering into a contract is just like marriage! Ironically, divorce is much easier than getting out of some contracts. Get to know the people or organizations you intend to work for/with. If you do not understand your client and their expectations, it is highly unlikely you will satisfy them.

Know thy limits…

In early 2003, I had been working in public accounting and dabbling in consulting with contractors. I got a reasonably good understanding of controls needed to complete a job and started feeling like I could handle doing a project. I mean, how hard could it possibly be, right?

My father-in-law worked on the Navy base as a contractor, so I figured I would get some advice from him. He gave me the general manner in which to pursue projects and some preliminary information as to how to get work on the base. Hindsight is twenty-twenty, but I sure wish someone had explained to me what I am about to tell you.

Unfortunately, at the mature age of 27, I knew a lot. I decided to go ahead and see if I could get some work for myself. I started making calls and networking, and before long, I had an appointment on base with a large prime contractor performing under a MATOC (contract with task orders). The meeting went great. We hit it off; I scanned his office figured out what football teams he liked, and within a few minutes, you would have thought we were best friends. At the end of our meeting he said, "I like working with people that I connect with, I will keep you in mind for the next project."

If memory serves, about two weeks after that meeting the phone rang and on the other line was my new friend. After a few niceties, he said, "does your company know anything about coded vessels?" Without hesitation, I said, "absolutely!" Great, come by my office and pick up the information to get this job done.

After disconnecting, I thought to myself: "What is a coded vessel?" Thank God for the internet! I searched and quickly learned I had just agreed to install a huge water storage container with all kinds of detailed specifications and hygiene challenges. I drove to the base the next day, picked up the contract information, and began to act like a coded vessel expert. I think I even answered a few questions while getting the packet.

I had no idea what to do next or even how to find a coded vessel. Fortunately, my father-in-law graciously stepped in and handled the project. If not for him, I might not have ever done any more contracting. Quite frankly, I did not understand my limitations, nor did I care at the time. I just wanted a job!

All these years later, I see this same scenario take place almost daily. Contractors pursuing jobs not aligned with their capabilities or going after work clearly, not part of their expertise. If the project is overly technical, and you do not have the right personnel, pass it up. Never take on a project with the thinking that you will learn as you go- it does not work. Stay focused on what you do know how to do or what your limits allow you to accomplish. That is not to say you should never push yourself, but boundaries need to be approached slowly, not aggressive or irresponsible acceleration.

As an accountant, I would be remiss not to talk a little bit about finances when discussing limits. In economics, they refer to limited resources as *scarcity,* meaning a lack of necessary resources to accomplish the objective. One point many contractors forget to consider when evaluating limits is the cost of preparing a bid or proposal. Before you even begin the proposal process there should be an internal calculation as to what it will cost for the acquisition. Don't pursue a proposal if you do not have some idea of what it will cost if you lose.

In 2008, I contracted to perform my favorite of all consulting task—cost management. The client had recently been awarded about ten million dollars in sole source work and at the same time had developed another relationship with a small base to perform small jobs totaling about one hundred-thousand dollars per project. After a few weeks and some skillful negotiations with the owners, I had them agree to let me set up an entire system for tracking project performance from the beginning to the end. After setting up the big jobs, I began to look at these little jobs to see how they performed.

The project looked great, but I noticed in their overhead the estimating cost for the last period seemed very high. At first, I thought this might be due to the larger projects, but I needed to confirm. After scrolling through the account, I noticed that the estimating cost mainly came from the smaller projects. I then met with the estimator and began asking questions about the bidding process, the smaller jobs, and budgeting. Unfortunately, he did not understand the idea of a budget and seemed more concerned with keeping his job than assisting me in identifying the problem.

After this runaround, I asked the owner if I could review the RFP for some of the small projects. What I found next blew my mind. Every small job they produced required a two-volume proposal with a technical and management approach and a fully cost-loaded schedule for uploading to RMS (software program for the Corps). They outsourced the proposal writing and the schedule preparation and completed the estimate in-house. They were spending just a little over what they made to prepare for these little jobs. No one took the time to evaluate the cost of the proposal; therefore, these little projects slowly consumed their limited resource - cash.

Just to finish the story, they took this analysis to the government and explained the situation. The representative apologized and said, "I wish you guys had pointed that out earlier because those requirements are just too burdensome for these small projects." If only someone had taken a few extra minutes to understand the cost of the bid, they might have avoided losing on the first few small projects.

In today's market, you absolutely must analyze the cost of everything you do! Make sure you set up a budget at the beginning of the year for the estimating department even if only one person does the task. Hold the people in that department accountable to that number by making them learn what your actual costs are in pursuing a project. Obviously, this information is crucial when deciding on the go/no go. Know your limits!

Read, read, and then read some more...

Contracting is very difficult. There are thousands of pages of specifications, codes, drawings, and opinions that require due diligence. If that is not enough, every regulatory agency imaginable is also watching everything you do or don't do. Wouldn't it be nice if you all you had to do was worry about one rule?

When my children were younger, I thoroughly enjoyed vacations spent aboard a cruise ship. Once we arrived on the ship and went through the safety briefings, the kids would disappear, and my wife and I could relax. The best part of the whole trip from the kids' standpoint was my one simple rule, "Every night be at dinner, dressed appropriately, with a good attitude." They loved this rule because, in their little minds, they were free the rest of the time. My youngest once said to me, "Living with only one rule, ...wow."

Good news! I am about to give you the one rule, the most critical aspect of the entire go/no go process...read, read, and then read some more! If you put this book down right now and you learn this one simple rule, then you will be way ahead of most of the people in the construction industry. The number one reason contractors fail on projects is their lack of reading what is required. They never read the contracts, the specifications, plans, or any of the information provided and show up on the job hoping for the best.

While writing this book, I am at an Air Force base in Northern California on a project. Before starting the project, I carefully reviewed all the documents, specifications, schedules, plans, California codes, and base-specific information. It took me several days to review all of the documentation. I then prepared quality, safety, and environmental plans specific to the project. I incorporated all of this information into subcontract agreements and began disbursing all of this information to the various subcontractors I selected to work with on the project.

After hours of reviewing the documentation, I prepared and sent all of the information to the various subcontractors. I included all of the documents I had studied and a sixty-page subcontract that covered everything from payroll to specifications. Within fifteen minutes of my sending the documentation, one of the subcontractors sent back an executed copy of the agreement. I made a mental note for the superintendent because I knew we would have issues with this subcontractor at some point during the execution of the project.

About one month later, my superintendent requested I attend the job due to some serious issues with one of the subcontractors. Would you like to guess which subcontractor? I got to the job trailer and listened for fifteen minutes while this subcontractor told me that he did not bid to pay his guys such a high rate, he never agreed to or would agree to all of the safety measures required and because he did not agree to all of this nonsense he wanted a change order. Finally, he let me talk. I simply confirmed his name and compared it to the information on the subcontract agreement. I then asked if he initialed all of the pages of the subcontract agreement and if he signed the agreement next to where the notary signed. He never read any of the documents, nor did he even realize what he signed. You have to read everything.

One more illustration because I really hope you understand the importance of this concept: A company I worked with recently employed two seasoned professionals, a project manager and an aspiring estimator. The estimator had just completed a career as a superintendent and hoped to transition into the office to eliminate traveling. The project manager recently retired from a very large contractor that exclusively serviced the government. The point is both individuals should have a solid understanding of contracting with the government and reading documents.

After a strenuous review of the RFP, the management, along with these two new team members, decided to pursue the project. Before beginning the estimating process, the owner requested a brief training session for review with the new personnel. Understanding the importance of reading, the entire session addressed the importance of reading all the specifications. Both gentlemen wholeheartedly agreed with the logic and allegedly spent hours studying the various documents.

As the bid due date drew closer, the bid manager began inquiring about RFI's and responses. The bid manager did not question them because she doubted them, but rather she wanted to confirm their acknowledgment of key changes resulting from the questions. When she finished her question both high-paid, seasoned individuals had what I call the "deer in the headlights" look. I certainly hope I don't offend any animal lovers, but when you are from the south you become familiar with this look because it is what you see when you slam on brakes hoping not to slam into the deer. After a few seconds, the bolder of the two asked, "Where is this information?"

Ironically, on the second page of the RFP in bold letters the Corps of Engineers outlined the exact procedure for reviewing and submitting RFI's. Here were two highly trained and compensated individuals potentially bidding a job incorrectly simply based on a concept we all learn before fifth grade—reading. Just so you know, when you submit questions to the government during the request for information period, make sure you are not asking questions that are right in front of you in bold ink!

A successful decision process requires a time investment. If you pursue a project without a full understanding of expectations and requirements, the risk is much greater than the potential return. Also, reading assists in developing a strategy that becomes very important later in the bid process. Don't make a poor decision based on not using the one important skill fifth grade provided.

Success or failure on projects is predicated on the amount of time one familiarizes oneself with the specifics of a project. Read, read, and then read some more...

At the end of the day, your team should be more familiar with the RFP than even the people who wrote it!

Attend the site visit...

In my opinion, there are two approaches for project acquisition from the standpoint of contractors: the statistical approach and the calculated approach. The statistical approach logic says, if I pursue enough projects, I am bound to win a few. I know contractors that bid 50 million a year with the expectation of only winning 5 million. On the other hand, the calculated approach requires the contractor to pursue a select number of projects based on their understanding of the project and the alignment with resources.

The second approach involves contractors that pursue a high kill rate (amount won/amount pursued). In the statistical approach, the ten percent requires a lot of bidding, but in the calculated approach the kill rate may be as high as seventy percent meaning if a contractor pursues work there is a seventy percent chance they win. Keep this in mind, we will discuss later.

It never ceases to amaze me how many contractors bid on projects without taking advantage of the site visit. I understand the rationale behind skipping a visit if you approach bidding statistically, but if you use the calculated method a site visit is a necessity. If nothing else, visiting the site enables the firm to get an early understanding of the work environment, the people, and the potential subcontractors and vendors. Personally, attending a site visit enables me to begin strategy development early in the process, assisting with developing the approach to a victory.

Before we leave this subject, a few tips may make your visit more advantageous. First, never attend a site visit without a solid plan of attack. What does this mean? Make sure you study the available documents prior to the walk, write down questions, and then attempt to find the answers once on-site. Second, generate a list of subcontractors or suppliers you might need for the project and either invite them to attend or identify ones already at the walk. Third, scope out the people who attend and listen to find out if there is an incumbent—a little due diligence at this stage goes a long way. Finally, before you go, write down what pictures are needed. For example, if the job requires demolishing a shed, write down the required shots you need to assist with bidding. Sometimes, it might help to ask one of your technical folks about what they might want to see or know more about. Ultimately, it comes down to planning the job walk, not just showing up and looking pretty! Do not attend a job site visit without a plan!

Another tip, when visiting a remote job site stay at the location for a day or two after the visit. This gives you an opportunity to understand the environment in which you must perform. For example, how close are resources, such as supply houses, subcontractors, and equipment rental? Also, I go and visit subcontractors and speak with them regarding the project. You can learn a lot by visiting the office of a subcontractor, and when you speak with them in the future there is a name with a face. Spending just one extra day on location is well worth the extra time and dollars. Remember, if you approach projects in a calculated sense, it is worth investing whatever necessary to increase the probability of victory. The more you know, the better decisions you will make!

Make it a math problem…

During my college days, I continually had one question surface in my developing mind, "Will what I learn today have any practical application for my future?" About 60 percent of the material I learned failed this test; however, I remember two subjects that I knew might be of value no matter what career I pursued: statistics and quantitative analysis. These classes posed the most significant study challenges, but the practicality made both subjects become my favorite.

When I wrote the first version of this book, I lived in beautiful Las Vegas. Ironically, Vegas deals in statistics, probabilities, and math more than anywhere else in the world. Gambling is a science if you do not believe me do a little research on the analytics used in casinos. I am not a gambler, meaning I do not walk into the casino and throw money on the tables, yet I do invest in the stock market, which is almost the same. You ever heard the expression, "the house always wins"? The house always wins because the house does its homework and understands probabilities, statistics, and how to use data to understand everything you or I might do while at the casino. Casino's make money, isn't that why you are in business as well? As a contractor, you should do your best to always be the house, understanding data, making decisions based on math, and always winning!

In Vegas, it all about the odds. If the odds are 9 to 1, this means there is a ten-percent chance of winning (1/ (9+1). Pursuing a project with odds of only ten percent seem a little foolish yet think of all the contractors that do this on a daily basis. This might be a great time to learn a little something from the gambling community.

When deciding to pursue a project, you must make sure the odds are in your favor. What are the chances or the probability of you winning a project you are pursuing? What variables do you need to consider when determining the likelihood of victory? For example, if you are bidding a job in your hometown, with an owner that you worked for previously, that loves you does that increase the probability of winning? What if you are pursuing a project in a remote location for an owner you do not know, doing work you are not familiar with, and it requires you self-perform a significant portion of the work, which is not something you usually do? What is the probability of winning in that scenario?

When working with probabilities, it is imperative to consider data relevant to the outcome. Make sure the components of the chart relate to past negative or positive results. For example, if a company only bids projects in one city and does not work in other locations, then the tab regarding project location is not relevant.

	Weight	Score	Probability	%%
Competitive Advantage	30	5	1	0.30
5-Is there a clear path to victory identified? 4-Do you have a strategic partner involved? 3-Do you have an unique efficient project approach plan? 2-Are you just bidding to bid? 1-Is this a pricing battle?				
Project Set Aside	10	1	0.2	0.02
5-8(a) Sole Source/Direct Work 4-8(a) Competitive 3-Woman Owned 2-Hubzone 1-Small Business				
Past Performance	6	5	1	0.06
5-Self performance on previous projects identical 4-Similar experience, not exact 3-minimal correlation 2-very little connection between past performance 1-No similarity				
Agency Familarity	5	4	0.8	0.04
5-worked with the agency at this location previously 4-worked with agency but not at this location 3-familiar with agency but no work performed 2-worked with similar agency or personnel 1-never worked for the agency				
Project Location	14	3	0.6	0.08
5-location is in area of previous construction or it is in home town 4-location is in large area with abundant resources (suppliers/subcontractors) 3-location is in remote area a minimum of one hour from a significant city 2-location is extremely remote...a minimum of two hours from any city or town. 1-location is non-issue				
Preparation time available	10	5	1	0.10
5-Two weeks or greater (Ten working days - no other jobs bidding) 4-Two weeks or greater (Other jobs in process) 3-Less than ten days 2-Less than eight working days 1-Less than one week				
Job Walk Attendance	15	1	0.2	0.03
5-Did someone from the company attend the site walk? 4-Did a subcontractor worked with previously attend the site? 3-Did someone walk the job bidding with us exclusively? 2-No one attended the site, but it is a ground up project? 1-No site visit.				
Project Value	10	5	1	0.10
5-The projected value is within the relevant range for bonding? 4-The project is in the relevant range; however, cost to complete is greater than 50%. 3-The project is in the relevant range; however, cost to complete is greater than 60%. 2-The project is in the relevant range; however, cost to complete is greater than 70%. 2-The project is in the relevant range; however, cost to complete is greater than 80%.				
Totals	**100**			**73%**

I wrote this book to make things simple, so do not fear this is not a hard calculation. I will tell you, this formula I have used for years to make decisions on projects, and it has been an excellent resource for making the right decisions on which projects to go after.

The weighted average column represents the importance each category represents. In the illustration, one might observe the high weight assigned to a strategic plan. This simply means management's main concern is strategic planning. Each category includes five different choices. These choices then convert to percentage points. The company decides what percent determines a go/no go.

Normally, in my experience, I do not like to pursue projects if they do not score at least seventy percent. If there is absolutely no work on the books, I might lower the percentage; however, if I am really busy I might increase the percentage to as high as eighty or ninety percent. This number is adjusted based on the economic environment of the company. If nothing else, this formula requires one at minimum to consider the risk versus return on the project and make an informed decision.

In economics, scarcity is the term representing a shortage of resources. Every firm, no matter the size, deals with this reality. When you hear the word economics or scarcity, you immediately think of money, but that is not always the case. Sometimes you do not have the personnel to bid all the released jobs, nor the time. Unfortunately, clients do not really care much about your level of resources and release lots of great projects all at once. For as long as I can remember, each September, government contractors experience this concept firsthand. The government ends its year on 30 September, meaning they need to spend any leftover money or lose it! In some cases, contractors literally pick and choose between the various projects hoping at minimum to pick up some work for the next year. Unfortunately, in some cases, contractors attempt to bid too many or simply choose projects not suited for their capabilities. The reason is they do not incorporate a procedure to evaluate projects based on what is best for their firm. Yet another great reason for this tool. By incorporating math, one can compare two separate projects and decide which to pursue.

There are a lot of other great math tools for making decisions that are beyond the scope of this book; however, I would recommend you do some additional research because the more you evaluate projects the better you will get a selecting what is right for your firm. Also, it is important to consider the financial implication related to selecting a project, such as margins, cost, etc.

Whether one incorporates my method or creates another model, tools must be in place to approach decisions mathematically. The model I provided is a good starting point for most firms and is sufficient to assist with making good decisions. The next few paragraphs present some generic categories for consideration. This tool is formatted to meet the needs of your internal environment.

Competitive advantage: Arguably, this concept is more important than any other factor in the evaluation process. Unfortunately, many contractors pursue projects for no other reason than they need work to remain open! If a company simply pursues projects with no competitive strategy in mind, the success rate drops significantly. Also, from a long-term standpoint, companies normally fade away when the economy slows down if they do not create a competitive strategy and live by it. If there is no clear path to victory on a consistent basis, it is simply a matter of time before dissolution.

If you are in this industry for any length of time, you quickly become aware of how rapidly companies come and go. You see companies so busy that it seems as if they are in great shape, yet a year or two later they are out of business or they hang on barely. To be honest, in my career I have owned a few companies just like this, and when I look back there is one reason for this type of performance—lack of a clear, competitive strategy that maintained a competitive advantage over my peers. The last thing you want to do in this industry is become a commodity.

Business terms seem to become prominent within an era. For example, not too many years ago the buzzword *paradigm shift* revolutionized the business world. One could hardly attend a seminar without hearing that word. Now, the word is *strategy.* It starts to lose its significance after hearing it a million times; however, that does not change its importance. Strategy is best defined as "…the unique value propositions a company offers its customers. Competitive advantage lies in the activities, in choosing to perform activities differently, or to perform different activities from rivals" (Magretta 2012, p. 174).

When considering pursuit of a project the question of *competitive advantage* demands attention. What value does a company offer that is unique? More specifically, what makes a company unique or different from its competitors on this project? Why should one company recieve this award over the other? If you cannot answer this question during the review phase of the project, most likely this is not the project for you. If you continue to pursue projects of this nature, you become a commodity and slowly but surely you run out of resources and end up shutting your doors. Competitive advantage is so important to the longevity of a company.

Early in my career, I had the opportunity to work for an older gentleman who successfully started a car dealership and built it into a thriving operation. As a matter of fact, it seemed everything he touched turned into gold over his sixty-year career. A few days after I started working with him, he set up a meeting to discuss the specifics of my job and his expectations.

If you know me well, you know listening to percieved nonimportant material for long periods of time is not my strong suit. That is a very nice way of telling you I have severe ADD!!! He went over company policies and all of the required discussions, but all I could think about is how he had built such an empire. Finally, we came to the end and he asked if I had any questions. Politely, I asked if it would be okay if I asked him some personal questions about his career. He obliged.

As I spoke with him it seemed like he offered nothing different than anyone else, but he had built this great company and engaged in so many other successful ventures there had to be something I was missing. I blurted out, "What makes you so successful?" I sit back waiting for some philosophical answer that would change the world. His response, "Make sure to do stuff no one else is doing." Then I asked how he chooses such excellent investments; it seems everything he touches turns to gold. He responded, "When I decide to invest in a business I look for one thing: what do they do that no one else does like them?"

Frustrated with the simplicity of his answers, I decided to probe further, "What is the one thing I can do in my career to stand out?" His response: "Whatever everyone else is doing, do the opposite, be different, be unique." Immediately following that someone interrupted our conversation, and he looked at me and said, "Success is not that complicated, really its not. Just find what makes you unique and don't change!" At the time, I discarded his comments, but in reality there is no better advice.

Unfortunately, I never had the opportunity to really talk with him in that format again. He passed away the following year, but a few years later his advice helped me start my first company. My first company began with what I now refer to as a "nitch" (yes, I know niche is the proper spelling). Yes, that is a southern term, and I am going to explain it in detail so everytime you think of strategy the word "nitch" will come to mind.

The first business I started catered to companies with large fleets of vehicles. I set up some simple software and managed all of the maintenance schedules for the fleets, telling the owners when vehicles needed servicing, including tires. Then I would go and negotiate purchase deals with large suppliers that provided deep discounts on parts bought in bulk. I worked with large fleets that had hundreds of vehicles and spent millions a year on repairs and maintenance.

As the business grew, I decided to also provide maintenance services to the fleets. I hired individuals to visit the fleet offices and service the vehicles after-hours, eliminating downtime for companies. This service really did well because it assisted companies with inefficiencies, and it kept their money-making vehicles on the road.

I started interviewing for potential technicians, and after about four boring interviews I met an older gentlemen who seemed as if he had just stepped out of a *Beverly Hillbillies* remake. Despite his appearance and demeanor, I actually liked the guy and started talking with him at length about his past experience and his capabilities. I remember asking for his resume, and he handed me a folded-up paper with a handwritten account of his work history. Needless to say, the grammar and the writing did not qualify him for an office position, nor did I feel comfortable at this point with hiring him to do anything.

I figured I would ask him some standard questions and then give him the old, "I will call you after interviewing all of the candidates." I led with my typical question, "Why do you want to work for this company?." I will never forget his answer. He said, "Sir, the reason I'm here today is because I think you found yourself a 'nitch.' Do you know what a nitch is, sir? A nitch is similar to an itch, but it is something everyone wants versus an itch is something everyone wants to get rid of. This business idea is a nitch, and I want to be a part of it." I hired him right after he told me this, and he worked for me for a long time. He might not have had the best writing skills, but he certainly understood business a lot better than most.

What is your "nitch"? What do you offer that clients need/want? What is your competitive advantage on this particular project? Do you have a defined nitch that makes you special or unique? That is what you must figure out in order to truly understand if this is a project worth pursuing.

Make the decision...

When it comes to decisions, there are thousands of books written on what to do when at a crossroads! Every speaker has some unique approach to making decisions, a step- by-step process, or the good old trust-your-gut approach. Unfortunately, a lot of us have known our gut for a few years, and we are a little scared to depend solely on its direction. On the other hand, the gut provides a lot of experience and knowledge that is worth considering, and it is the only advice that contains no outside bias. I think you should trust your gut, but only under the conditions explained in the paragraphs below.

Typically, decisions made strictly by the gut are backed up with a lot of data and math, but it sounds more exciting and makes people seem smarter when they just go with the gut! Let me explain the use of the "gut" in the decision process by illustrating with a story you are probably familiar with.

January 15, 2009, started out just like every other day for Captain "Sully" Sullenberger, a pilot for US Airways. Nothing special or unique about the day or the flight, but within three minutes after takeoff, his life changed forever. The plane flew right in the path of birds disabling the engines and forcing an emergency landing in the Hudson River. His expertise allowed him to safely land the Airbus 320 in the Hudson River, saving the lives of all 155 passengers and crew.

Most people remember the heroic landing, but few know what transpired in the weeks after the accident. While the media portrayed him as a hero, citing the events as the "Miracle on the Hudson," the National Transportation Safety Board and his employer questioned his decision to land the plane in the river.

The NTSB, Airbus, and US Airways did not consider Sully a hero but rather a pilot who ignored the onboard computer and made a "gut choice" that potentially put people's lives at risk. The various investigative agencies executed computer models concluding that the plane would have landed safety at the airport if Mr. Sullenberger had followed the computer's instructions. It became obvious the agencies simply thought this pilot had made a risky decision based on his gut instead of depending on the controls in place. As a matter of fact, twenty-two different simulations proved this theory!

The movie about this story, of course, included a little more drama, but surprisingly it re-created the situation rather accurately. It even illustrated the press portrayal of him as a hero while the investigation by the authorities showed a rogue pilot making the wrong decision.

What really makes the movie incredible is the ending where they reenact the final hearing on the crash. Mr. Sullenberger, the crew, and investigators are all in the same room listening to the findings from the various simulations. The evidence seems overwhelming, and despite the strong affection held by those present for this seasoned pilot, the doubt in his decision mounts.

There is a brief hesitation, silence in the room, and then Mr. Sullenberger proceeds to explain not only why he made his decision but also why the computer models are incorrect. They make the minor changes and consider the variables he presents, and the room is astonished as the computer models start showing a simulation ending with a crash. Within a few quick seconds, under pressure, Mr. Sullenberger made multiple calculations in his head and landed the plane safely in the water. Truly amazing!

Before he starts explaining his thought process, it is easy to assume this experienced guy did make a gut decision. Midway through his explanation, I realized something. He did make a gut decision. Unfortunately, we don't really use the term "gut decision" correctly. Mr. Sullenberger logged just over 19,000 hours in his career, worked on multiple crash investigation teams, and trained pilots in the military. He spent an average three hours a day, every day for thirty years in the cockpit of a plane. He had a very healthy gut! If you invest this time into understanding RFP's, you then will also have a gut that is trustworthy.

What is the takeaway? Simple! Read the RFP, do the homework, spend the time necessary to thoroughly understand all the items discussed previously, and then make a "gut" decision on whether to pursue a project.

No Go

As an entrepreneur, one of the most difficult things to do is end up with a "no go." A "no-go" simply means that, based on all the analysis, the project in question is not the right fit for your company. In other words, this is not one worth pursuing. Don't waste any more money on research; just let it go. It belongs to someone else. This might seem a little frustrating at first, but if a company commits to this process, they will slowly watch their kill rate (projects pursued versus projects awarded) increase to the point they only pursue work they ultimately win.

In my younger days, I envisioned myself operating heavy equipment as a career. Daily I loaded up my Tonka Toys and headed off to the backyard to develop roads, neighborhoods, and retention ponds. Naturally, based on my level of experience, when I became of age, I purchased some real heavy equipment and started a business venture doing dirt work, land clearing, and demolition. I absolutely loved the work, probably more than anything else I have done in my entire adult life. In 2007, Hurricane Katrina, the costliest hurricane on record at that time, devastated New Orleans and destroyed entire subdivisions and communities within a matter of hours. After the dust settled, FEMA released contracts to destroy severely damaged homes. With an excavator, two Bobcats, and a crew of five, we left for New Orleans.

After the briefing and the assignments of work, my crew and I headed off to destroy the first few houses. Despite their claims, organization is not a strong suit with most government agencies; therefore, instead of doing one street of demolition, we had to move equipment from one neighborhood to another to do our work. While destroying one building, your competitor might be right next door doing another building at the same exact time. On our third house, we worked directly with a competitor starting demolition at the exact same time. He destroyed his building in about half the time it took us to destroy our building. After he finished, I stopped and went over to him to find out what he did differently.

After a few minutes of chatter, I asked, “How did you tear that building down so quickly?” The response he gave has stuck with me all these years. “Operating an excavator is just like doing anything else. If you want to be quick or get the most of out of something, you must learn to use every movement to your advantage. Most operators waste a lot of movements and end up spending more time on the work.”

When you choose not to pursue a project, do not waste this movement. Perform an internal debrief on what made this project unappealing. Learn from each movement and continually hone the internal skills of selecting projects. Determine what weaknesses prevented you from pursuing the project or how you might improve to pursue a similar project in the future. On the other hand, maybe solidify your position and create policies in the future to ignore projects of this nature. The point: use every movement to your advantage.

Go

Congratulations! All the research, calculations, and gut feelings led you to the most exciting part of contracting—the Go!!! It paid off, and now with a green light the real fun begins! No matter how long I do this, I always feel like it is Christmas morning when we make a decision to pursue a project! Now it is time to do the one thing everyone talks about but very few do—planning.

Of all the books, articles, television shows, and every other thing that has flashed before my eyes over the last forty-five years, there is one quote from Stephen Covey that in my opinion sums up the most important element of any business operation and, more specifically, contracting: "Begin with the end in mind…all things are created twice. There's a mental or first creation and a physical or second creation to all things" (Covey 2004, 98).

To me, beginning with the end in mind means taking the time to know where you want to end up and then moving in that general direction. Set goals for the project. What do you want for margins? How do you want to satisfy the client? Maybe you want some specific training while on this project, or maybe you want to work on self-performing. Whatever it is, if you don't apply this principle, how will you measure the success of the project? Remember, success is simply reaching the goals you set. Once you set those goals, the plan (strategy) is what takes you from where you stand to where you want to be at some point in the future.

There is an old rule out there that I stressed to my team over the years: $1, $10, $100. The concept is rather simple. If we do appropriate planning prior to the bid, it costs us approximately $1.00 per hour. If we forego that planning until after award, it costs us $10.00 per hour. And if we forego planning all the way to the start of the project, we pay $100 per hour. The point is simple. The bidding stage is much more than just throwing together a few numbers; it is our promise to the potential client that there is a thorough understanding of the project, and we intend on executing accordingly.

One final note on the importance of planning. Planning does not guarantee success; however, as the British businessmen Harvey Jones once said, "Planning is an unnatural process; it is much more fun to do something. The nicest thing about planning is that failing comes as a complete surprise, rather than being preceded by a period of worry and depression." If nothing else, avoid the worry and depression!!!

The next few chapters provide step-by-step instructions on how to take the potential project from an estimate to a deliverable. If you apply a fraction of what is covered in the following text, I assure you that business is going to improve, and for once you will control the direction of your company, increase margins, and make contracting a viable business. This works for any size operation or for a contractor or subcontractor.

As you consider the information moving forward, refer to my opening illustration to decide the type of estimator you want to be or the type of estimating department you desire to create. If you are content with just pursuing projects with no purpose, tacking on contingencies and hoping for the best, this is not for you. On the other hand, if you want to create an operation built on accuracy, driven by attention to detail, and that is strategically motivated, keep reading! If you want to be the type of estimator who delivers a promise to customers instead of an educated guess, this is the right process for you!

Many read this information and respond with, "What we are doing works, and we make a lot of money." That thought may very well be true; however, as everyone would agree, construction is becoming more and more difficult as regulations and codes become more prevalent. Not to mention, big labor and requirements for individual certification change the playing field as it relates to finances. The point is the old way is not going to stay for long. Clients are raising expectations in this industry as well, meaning we must prepare and respond in order to remain a going concern.

One last thing: as with any change, there are hurdles! As you implement these processes, estimators and project managers alike might argue to keep the old systems in place. Some might even go to great lengths to shoot holes in the process and prove that their current system works better. The truth is the current system works poorly, which is proven every time a new project begins and the people in the field try to implement what they think the estimators meant. Then they complain and create a whole new approach to the project. Yep, that system is not working. A project is simply the management of both time and money. These two components, appropriately balanced, are what create a successful project. Don't let these people change your mind. Focus on the type of estimator you desire to be.

CHAPTER 2

Step 1: Create the Team

Isn't it funny what memories you keep from your childhood? As a matter of fact, it is weird to me what memories I hold from just five years ago. My wife has worked as a Mary Kay beauty consultant for years, and she absolutely loves it. When she first started, I attended several events with her and observed how the business operated. To be honest, I was impressed. I decided to do some research on the founder, Mary Kay and began reading anything I could find about her and the business. In one of her books, she said the following, "People don't remember what you say, but rather how you make them feel." I think that explains why we remember certain parts of life and forget others. Our memories hold on to the parts of life that impact us emotionally.

With that said, to this day, I vividly remember an event occurring almost daily in my small neighborhood in Lynchburg, VA. In the summer, every neighborhood kid came out to play, and eventually, someone would recommend splitting up into teams to compete in some sport. The older kids led the charge and would pick the teams. Us smaller kids, huddled together hoping to get picked, and most importantly, not be the last kid selected. If you were picked last, it meant you were not good at the sport, or you were not one of the cool kids. Either way, getting picked last sucked!

Most of the time, the older kids picked basketball as the preferred sport. At the time, I had some issues with my vertical stature, little guy, and I really did not do very good at basketball considering my height disadvantage. As

I write this, my feelings from those days are surfacing! Anyway, I remember standing in the crowd of small kids, hoping and praying I would not be the last kid standing.

Fortunately, this old school method of selection does not apply professionally, or at least it does not happen publicly. The objective in any game is to win, winning requires putting together the right team. When the players on the possess the necessary skills, the coach puts the right people in the right spot, good things happen. It all starts with making sure the right people are on the team.

When it comes to winning construction projects, the principle is still relevant, the right team must be in place. Selecting the right team requires a manager to understand the project and identify the skills of each player to put together the best bid. Before we get into the process of bidding, mentally, line up your folks and think about who gives you the best chance of winning.

Before I start, let me say not every project requires a full team. Smaller projects may not demand as many folks, or in some cases, you may not have the staff to fill every position. It is okay. As long as you understand the roles of each person. If the project is large, then you simply provide a team under each of these people to carry out the various duties.

Bid Manager – First Position

The longer I am in business, the more I understand the importance of management. Processes, businesses, academics, programs, and individuals fail without oversight. Bidding is no exception; it must be managed. A successful construction company is like a three-legged stool, with each leg being equally as important as the others (Leg1: Getting the work, Leg 2: Doing the job, and Leg 3: keeping score). Despite their equal importance, the estimating/bidding leg contains the most severe consequences and can put a company out of business quickly. Manage the bid process!

The bid manager is the key person in the acquisition process. They are the "Big D"—the decision-maker for all things related to this proposal.

Their responsibility is to oversee and execute the process. Many do not understand the importance of management and tend to assign a manager and then add a great deal of responsibilities or keep them doing other work during the procurement process. For this to work, let this person do nothing but focus on winning the project. This is the person fully responsible for successful acquisition. *Mastering Rockefeller Habits* gives the following advice: "All projects, line items on an income statement, priorities, and processes must ultimately be owned by a single person, even though there might be hundreds of people who have some kind of sub-accountabilities and responsibilities in seeing something completed" (Harnish 2002, 97).

As an owner or manager of a construction company, it is imperative you understand the importance of the bid manager. The bid manager is not only responsible for the bid, but he/she also is the crucial link between connecting the great plan devised in the office to actual execution in the field. If you do not think this link is crucial, then explain to me why jobs fail in the field. Surely, estimators did not comprise a bid package knowing it would fail. Nor do field personnel always screw things up, right. The reality is, many companies never transfer the strategic planning done in the estimating office into the field. You win the job with one strategy and perform it with an entirely different plan. Also, consider the money sacrificed if the field plan alters a vital element of the office plan. A bid manager eliminates this confusion and supports the transition from office to field, as you will see later.

Another mistake made when assigning this position is choosing someone who is a construction expert rather than a process expert or a manager. In most cases, the best bid manager is the least experienced in construction because this person is more likely to adhere to the process instead of defaulting to field experience. What you need is the person who is highly organized, motivated, and particularly skilled in management. Contractors tend to only hire people who possess technical skills and then wonder why they constantly endure business frustrations. If the person selected spends more time on technical construction issues than managing, you probably put the wrong person in the position.

A lot of companies miss the boat on hiring the right people. As a forensic accountant, I go into a lot of construction companies with issues because the office personnel consists of former field employees. On 9 April 1865, General Robert E. Lee surrendered to Ulysses S. Grant ending a tragic civil war with consequences still impacting society. The war began four years earlier, and by all accounts should have ended quickly based on the military prowess of the North, and its superior military capabilities, Four different northern general's failed to overcome Lee, despite their military expertise. Lee might not have been a better general, but he knew how to lead and manage people. The lesson here is that the right general is more important than a general with wartime experience! Make sure the bid manager is a general with the right skills for needs to be done.

Estimator – Second Position

As a general contractor, an entire team of estimators may be necessary to address s specific scope. Still, there should be one person on the team responsible for compiling all the numbers despite how many people actually participate in the preparation. The individual selected for this position needs to thoroughly understand the scope, organize all the information, and work closely with the other personnel to identify the specific activities requiring a cost. This is not customary; normally, the estimator works on his own and provides numbers at a pre-defined time. Not the case with this process. The estimator is fully engaged daily, making sure the estimates align with the overall strategy developed by the team.

This estimator answers directly to the bid manager. It is crucial that the bid manager stay involved with this individual to ensure he/she does not take shortcuts or make assumptions strictly based on experience. Also, the bid manager must maintain control over the estimating process to guarantee the estimator does not include contingencies or random charges to cover a deficiency in the bid process. For example, an estimator might receive a number that does not cover certain aspects of the specifications, but instead of doing a little more work to identify true cost he/she might just add a little extra to the number to compensate for the deficiency. This excludes the rest of the team and bid manager from understanding true cost for the scope. Any deficiency must be brought before the entire team to review.

Estimators typically do not like to share information, nor are they transparent with calculations. Sometimes, they act like a mad scientist with lengthy spreadsheets or intense calculations that no one understands or has the time to dissect. Asking a direct question turns into an hourlong explanation, which normally discourages further questioning. Do not let the estimator bully their way to running this process. Provide the estimator with the format for information and clearly define the expectations from the beginning. Most importantly, monitor all activity to ensure the information remains functional for the entire team's review. The objective for this individual is to identify cost for the various activities—nothing more, nothing less. The estimator's opinion is not included in the cost identification but later in the presentation of their findings to the entire team. This individual should make no decisions regarding cost without including the team.

Collectively the team created the project strategy (discussed in the next section); therefore, the estimator is not authorized to veer from this plan without bringing it to the attention of the bid manager and the team. For example, a strategy often includes self-performance, but an estimator does not want to do the takeoff; therefore, they obtain numbers from a subcontractor acting as if these numbers represent a takeoff. Make sure every calculation shows the work!

As an accountant, I learned a long time ago that skepticism is my friend. In college, my professors continually reminded us to show our work if we wanted credit for our answers during examinations. Sometimes, we came to the wrong conclusion, but we received credit for following the process. When I began working, my manager at the time further reinforced this point by continually asking us new accountants, "What state am I from?" We would all respond, "Missouri." Then he would say, "What is the mantra for the state?" We replied, "The Show Me state." Whenever we did not provide proper documentation to support a calculation or position, this would be the procedure. I learned fairly quickly to only depend on documented proof. Make sure your estimator (and the entire team, for that matter) understands what state you represent!

Make sure no one on the team, especially the estimator, works in a vacuum. Simply put, a lot of construction professionals like to keep information contained until the last minute, using their own special tools that no one understands. If you allow this type of secrecy, the team concept is useless, and you will not benefit from all the resources available to you on bid day. Make sure any tool utilized is something the entire team agrees upon using and is capable of accessing at any time. If a team member maintains a special spreadsheet for their calculations, make sure the entire team understands it and is capable of accessing it for review. Keep everything in the entire process open and ready for review at any time from any team member.

Another important idea related to working together is to make sure all documents are reconciled daily. The last thing you want is a document that is not up to date, and someone is working off information that is not correct. Utilize your software for checking documents in and out to ensure there is no outdated reporting.

Project Administrator – Third Position

This individual works as the support person for the team. They make a lof of phone calls, conduct specific research for the estimator or the bid manager, and maintain the Who? What? When? list (explained later). This individual also supports everyone on the team to ensure they receive communication about RFIs, changes in scope, and other information released from the client. The project administrator makes sure all documentation remains updated for each team member. I know the word "secretary" is taboo these days, but this really is what this person does during the process.

If the team is pursuing a government project, it most likely requires certain paperwork or documents submitted in addition to the pricing. The individual in this position prepares all of this information and informs the various team members what is needed in order to prepare for bid day.

This same person comes in every morning and checks the bidding website to make sure the owner did not post a modification or an amendment.

The project administrator performs a majority of the cold calls to subcontractors at the beginning of the bid process. The team prepares a script and tells this individual what trades are necessary. The project administrator makes multiple contacts, documenting every contact and passing this information on to the team for further review. I cannot stress the importance of this individual in the process. Do not assign this role to just any person in your office. Instead, select someone who is detail-oriented and understands the importance of what they do on a daily basis. Just throwing someone in this position is not an intelligent move.

Another crucial task administered by this person is the maintenance of minutes from the various meetings and conversations conducted during the bid phase. Every conversation or other form of communication is documented, ensuring all team members have access to everything related to the project and the subsequent bid. By the way, this person, if trained correctly, makes a great bid manager after they have a few jobs under their belt.

Project Manager/Scheduler – Fourth Position

Ideally, the individual filling this position is the same person who runs the job after award. A smaller project may utilize this person as the superintendent/project manager. Either way, this individual is going to be involved with this project after award. As a matter of fact, it is crucial this person be involved to make sure the strategic planning that went into bidding carries over directly to the actual deliverable.

As discussed previously, a key frustration noted in my years in construction is the lack of a successful transition between the estimating phase and the construction phase. Unfortunately, most of the time the plans carefully created in the office never make it to the field because the two do not communicate. If you worked in construction for only a brief period of time, you certainaly are familiar with the field blaming the office and the office blaming the field. From my experience, fault usually goes out to whomever is not present to defend themselves.

The primary objective for this position is preparation of a detailed project schedule. By far the most important part of the process is preparation of the schedule; therefore, the individual serving in this position must have a solid understanding of construction and the process for accomplishing specific scopes. Make sure you select the project manager with the most experience in the field most prevalent on the project. This is a defnite strength to consider during the go/no go phase: *"Do you have someone proficient in the scope of the work you are proposing?"*

After thoroughly reviewing the plans and specifications, this individual works closely with the team to prepare the WBS (work breakdown structure) and then sets up the definable features of work, the activities in each section, resourcing, and other key scheduling components. This is the key document in the process; therefore, it is imperative this individual is competent and is held accountable throughout the process by the bid manager.

The entire team utilizes the schedule throughout the bid process. In some cases they add lines and make changes, coordinating this with the project manager. They also provide input on activity times and the resources necessary to complete the task. After the first draft, the schedule becomes the one document everyone refers to in order to set up the project or assign resources.

The project manager is key in the next aspect of the process, developing the bid strategy. It is important they are involved in crafting the actual strategy for project completon, which should align with the overall bid strategy. The project manager also assists the estimator in communicating with subcontractors and reviewing and vetting potential subcontractors and their respective bids.

Earlier, I said the bid manager is the individual that ensures the strategy in the office makes it to the field. The bid manager oversees all aspects of the bid process ensuring each team member stays committed to the process including the project manager/scheduler that is eventually going to run the job in the field. The bid manager makes sure the strategic plan is rolled into the master plan (schedule) and the master plan is what drives the project after award.

The Wild Card

In twenty years of construction, I would say the "wild card" person has won more projects for me than any other individual. If this person is doing their job correctly, it is very easy for the rest of team to form a disdain for them quickly. In many cases, this individual is an executive of the company, an owner, a project manager, or another person not directly involved with the project but with some level of construction knowledge. In many cases, it is someone, as my Grandma used to say, with a lot of horse sense!

The wild card almost acts as an auditor. They listen to the rhetoric and then compare it to the actual documentation. Ironically, even though there are only about seven days lapsing from when a strategy is developed and when the bid releases, the strategy developed in the planning phase drastically changes. The wild card individual keeps everyone committed to the original plan and ask the questions that others miss due to their involvement. As an owner, I spent a great portion of my career serving in this capacity and recorded a lot of stories for this point. Make sure the person selected is a skeptical individual, preferrably with a little accounting background. After all, it is an audit!

Prior to ever starting my company, another company engaged me to more or less play this position. The owner requested I come in a day before the bid and go over everything with the team to make sure they missed nothing important. The project consisted mainly of demoliton and dirt work involving a lot of heavy equipment and removal of spoilage. Each team member proudly presented me their take on the work and explained the procurement strategy and the workflow on the project. The project budget came in around $10.2 million, and it looked like a great project.

The next two hours consisted of me listening to these well-trained, mature professionals explain everything from the schedule to the dollar values. As a matter of fact, they continued to talk so much it seemed impossible to do any analysis with them in the room. I took lots of notes and told them I would review everything and get back with them first thing in the morning.

I downloaded a copy of the schedule, which contained all of the resources, labor, equipment, and materials. The team did a great job identifying the various activities, resourcing them, and evaluating the risk. Each line item included the necessary equipment and the time frame it needed to be on-site. They also completed evaluations of the best rental times to reduce cost. The team collectively had over 150 years experience; they were a seasoned group of professionals.

After about three-minutes of review I noticed something that seemed interesting. Not only did they procure great prices on equipment, they also retained equipment that operated without a person. They did not include any operators to man the equipment for the project.

The next morning when I revealed my finding they all looked at me in disbelief. One of them argued for a few minutes until he saw it with his own eyes.

The wild card periodically reviews the project and uses common sense to identify issues. They also review all the documents submitted on bid day and ask tons of questions. The wild card is from the state of Missouri, "the Show Me state," and drills all participants, making them justify their logic.

Creating the right team is the foundation for successful projects. The right people committed to executing the process provide a winning combination.

Let me address some different audiences for a moment. First, the small business that does not have the resources to compile a team of this nature must use the resources currently in-house. For example, if your staff consists of only two people, divide up the duties between the two of you and follow the steps in the rest of the book. The point is to make sure the duties assumed are handled appropriately, and the result will end up the same.

I would highly recommend that large businesses maintain the same structure and just assign people underneath each key team member. If you utilize ten different estimators, make all of them answer to the lead estimator. I would caution you, however, to make sure there is only one bid manager—a "the buck stops here" kind of deal.

Not to get too far off of the subject, but the right people in the right positions is an absolute necessity for any organization to function successfully. "Those who build great companies understand that the utlimate throttle on growth for any great company is not markets, or technology, or competition, or products. It is one thing above all others: the ability to get and keep enough of the right people" (Collins 2001, 107).

CHAPTER 3

STEP 2: DEVELOP THE STRATEGY

When it comes to strategy and planning, my Dad wrote the book. Okay, not literally, but the man lived out the idea of strategy and planning long before it became the buzzword it is today. He plans out everything he does, even down to the smallest of details, which rolled over into our family vacations. When I was little, my Dad's office was in the basement. You opened up the basement door, walked down the stairs, and there he sat with maps, brochures, and notes all on his desk. He carefully studied all of the information and developed a detailed vacation plan that included every aspect of the trip down to where we stop for fuel as we drove to the destination. When I say he planned everything, I literally mean everything. He even planned the bathroom breaks based on the fuel consumption of the Ford Gran Torino, the family vehicle. All I can say, is thank God hybrid vehicles did not exist in my younger years!

Vacation planning began with his overall objectives for the trip. He invested quite a bit of time researching locations and determining what offered the most opportunities for his family. Don't forget, he did not have the internet, so the research he performed included reading brochures and making phone calls. Looking back, I am so thankful for his planning because even to this day, my vacations as an adult are never as good as the ones he executed.

Fast-forward twenty years to yours truly! Obviously, as a wise young man, I thought vacations should involve no planning, no structure, and no required activities. I looked back on my rigid holidays as a child and thought, I can make this better. Instead of making the trip stressful, why not just get in the car and go and do whatever comes to mind. After all, no plans, no stress is by definition, relaxation, right. The wife, four kids, and the gas guzzler headed to a relaxing, no pressure, no plans vacation.

The trip ended up in disaster both financially and personally. I spent so much money doing literally nothing, after a day or so the kids and wife started harping on me for not planning anything, and my last minute plans fell through because the places we visited had no tickets for sale or were booked for the day upon our arrival. On our ride back home, my stress level reached new heights, and all I could think about is my Dad!

Planning is at the core of everything we do personally and professionally. Even spontaneous people plan, they plan a block of time to be available, or they plan to be in a precise location with a particular group of people with hopes for a random event to occur. Planning is an absolute necessity! Unfortunately, it is not something we formally like to do. We talk about, we use it in speeches, but to sit down and make a plan is not an easy task.

When working with a contractor on a bid/proposal, one of the first questions I ask, "What is your bid strategy?" Typically, the look back at me with the deer in headlights look, or they say, "to win." Winning is not the strategy; it is the end goal. What they mean is, we do not have a bid strategy. We bid on a lot of construction jobs and based on statistics; we hope to win enough of the projects we bid to stay in business.

I want you to do me a favor right now. Pretend for a second, you are the lead estimating engineer for a $100 million construction company. The company after experiencing significant legal challenges is ready to emerge from bankruptcy and reposition itself in the market. There you are ready to lay out the acquisition strategy to put this company back on solid ground. Across from the table, sits a consultant named Douglas Allen excited to hear this strategic plan. Yea, you might be pretending, but this is a true story. The estimator, which for the record is a very intelligent individual, looks across the table at me and says, "we are going to bid $1 billion worth of projects this next year, and according to statistics we should win approximately ten-percent of them, which provides us with revenues of $100 million." Put yourself in my shoes for a second, my job is to evaluate this plan and present it to external shareholders as reasoning for providing continuing financial support. Would you feel confident presenting this method to very sophisticated outside shareholders? I know I didn't.

Since everyone loves statistics, let's review a few of them. According to the Bureau of Labor Statistics' approximately 20% of businesses fail in their first year, another 30% fail in the second year, and after five years another 50% fail. It gets better, after ten years 70% of businesses fail. Here is another great statistic, in construction only 35% of companies stay open longer than five years. Do I have your attention? If you are a contractor reading this, it is very likely you will not be in business very long. Unlike the ten-percent rule stated above, these are proven statistics needing attention.

There are literally hundreds of reasons why businesses fail, or maybe I should say there are literally hundreds of symptoms that we latch on to when businesses fail. For example, ran out of cash, poor bidding, could not find people, lack of payment from owners, and a host of other great symptoms for failure. Instead of talking about symptoms, why not address the reasons or the underlying problem that led to the symptoms and eventual failure. Here we go with my reasons for failure:

M	Missing Strategy (Business never has a plan – just show up and see what happens)
I	Implementation (Great plan never implemented – they chase whatever comes along
A	Alignment (they create a strategy that is unrealistic based on their core competencies)

Take a second and run any symptom through these three concepts and you will get my point. For illustration purposes, we will take the most popular symptom for failure and run it through my rational. Lack of cash is normally cited for the number one reason companies fail but is that really the case. Running out of cash does not just happen overnight, it typically takes several months or in some cases several years. Most of the time, when a contractor runs out of cash it is a surprise to them because they never did any type of planning as it relates to cash spending it whenever available. If they had a cash management strategy or plan in place, they would have realized the issue and possible been able to mitigate it before the woke up one day and realized they had absolutely nothing left. Take it a step further, some contractors do have a plan for managing cash, but then they go out and buy new equipment, overpay themselves, and then wonder where the cash went. Finally, contractors try and compete in areas they do not understand, or they will not get help, which either ends up in cash flying out of the business, or litigation, performance issues, etc.

For me, both personally and professionally, I can track my failures right back to one of my three points: MIA. Before you go all righteous on me, I do understand there are companies that legitimately fail for reasons outside of their control, but they are exceptions and I do not like making exceptions the basis for decisions. The point of all this is to understand what a key part planning is in business.

Now let's get back to the topic at hand. We started this conversation based on me asking contractors if they had a bid strategy in mind, right. Then I proceeded to explain to you why businesses fail, and now we make the connection. In construction, unlike other industries, every project is like its own little business. Think about it, you could literally set up every project with a different bank account, different personnel, and run it as its own business. When you pursue a project, you must think of it in this manner. How can you start a new business without a plan? What value proposition are you offering in your bid? How do you intend to deliver your services uniquely, which sets you apart from your competitors? Don't pursue projects based on the fact you are a contractor, pursue projects based on a unique approach that you know provides a competitive advantage.

What is a competitive advantage? How does that apply to bidding? Great questions... let's ask the foremost expert on strategy, Mr. Michael Porter. "Competitive advantage grows fundamentally out of value a firm is able to create for its buyers that exceeds the firm's cost of creating it" (Porter, 1985, p. 504). If you add value to the end-users expectations, then pricing is not the primary consideration. Follow the logic, if everything you bid is just done for the sake of doing it, then you become a commodity, which means it is all about the lowest price. You enter a price war, where you and your competitors work closely together to keep pushing prices lower, eroding margins. What are you really doing is competing to see which one of you will go out of business first. Once you enter this game, the margins keep going lower, you keep reducing cost, and before you know it, there is nothing left to work with the doors shut and the business ends.

We all know there is growing appetite out there to get contractors to do the work for almost nothing, how do we overcome? Great question. The answer seems simple, "Strategy is an internally consistent configuration of activities that distinguishes a firm from its rivals" (Porter, 1985, p. 334). It comes down to developing a bid strategy, a plan that sets you apart from your competitors and provides you with a competitive edge, or in simple terms, a higher margin than the other guy.

Let me give you a simple example from my contracting days. We were bidding on a small demolition project at Nellis, AFB, in Las Vegas, NV. I love demolition and really wanted the job. I sat down with my team, and after going through the process you are learning in this book, we decided to pursue the project. We shut the conference room door and spent hours trying to determine a strategic plan that would make us competitive. We ended up deciding to surround the property with containers instead of running dump trucks all day to the site. This would allow us to have one excavator on-site and get the containers dumped in the late afteroon, early morning, or on the weekend. After award, another contractor protested with the government saying there is no way they can complete this work with their pricing. About a year later, I ran into the contractor that filed the protest, and he told me his bid had a 1% margin. Guess what? No plan, no unique approach, just standard practice.

Don't worry, I am going to walk through every step of the strategy formulation for bidding projects and then give you some illustrations for bid strategies. This provides you the necessary tools to develop correct bid strategies.

Strategy Formulation Step 1: Familarize the Team with the Project

As an accountant, I have worked in a variety of industries. Despite the differences in each sector, there are also a lot of commonalities. For example, when I worked in construction every conference I attended, the speakers would spend a good portion of their presentation talking about how important it is actually to read the plans and specifications related to a project. When I visited a retail establishment, the owner complained about his personnel not reading the various reports generated from the system to assist in managing inventory. In my new career, the very first conference, the speaker spent the first ten minutes talking about how important it was to read the bond when working on a claim. Do you see a pattern here? People do not like to read - they make assumptions and ignore details. For this reason, my first step in developing a bid strategy is to spend time reading every document. The following paragraphs provide a little method I used in contracting and still use to this day whenever I am reviewing documents: PQ4R (Preview, Question, Read, Reflect, Review, and Recite).

Preview: When you first receive the documents, whether electronically or digital, take a few minutes to understand what is in front of you. Let your brain organize the information in a manner that will keep everything in an order that makes sense to you. This is different for everyone. If you are a die-hard contracting type, you might want to read the plans first, specifications, and then end with the RFQ. If you are like me, I like to understand the contract and then look at the other documents. Understand what each section of the papers represents and make yourself a little outline for reference. Many times, during this part of the review process, great questions will come to your mind, WRITE THEM DOWN! DOCUMENT YOUR BRAIN ACTIVITY! If I do not write stuff down during this phase, it goes away, and sometimes this is where your brain is still free and objective. Take advantage of the preview!

Question: You know what I used to tell my kids, "if you are not asking questions, then you are not thinking." If you read plans, specifications, and RFQ's do not generate one question, then stop what you are doing and go get a physical exam - you potentially could be brain dead!! Write down everything you can think about at this juncture. Maybe in your preview, you noticed a section on asbestos, and now you are thinking, "did they provide an asbestos survey?" In all seriousness, if you do not have tons of questions after doing the preview, stop, do something else, and come back to this later.

In some cases, specifications and contracts reference other documents, "clauses incorporated by reference." In Government contracts, sometimes they reference all types of FAR (Federal Acquisition Regulations), which they expect you to follow. I am amazed at how many contractors bid work and never review any of these referenced documents. Make sure you write down questions like, "What does FAR 42.202 require me to do, and what how does it impact bidding this project?

Out of all the things I see contractors do, not understanding what they are required to do is by far the number one thing that gets them in trouble. Here is the sad truth, if you do not know about a requirement, you cannot bid to cover the demand. Make sure you go through every aspect of the contract documents.

Read: Unfortunately, right here is where a lot of contractors/subcontractors stop. They do a quick preview, think of a few questions in their mind, and then start working on compiling the cost. Big mistake, which leads to significant oversight. READ, READ, READ! No, I mean literally pull up by the fire and read everything like it is a great novel. Highlight, take notes, answer the questions you created in the previous section, but most important, read every single page.

Estimators and Project Managers are notorious for skipping all of the contractual information and diving right into the specifications or plans. They figure the individual writing the proposal or someone in management will read all of the contract stuff, and they will get the numbers.

Several years back, my company pursued a large fencing project for all the Navy and Marine installations in Southern California. It only involved fencing, so the specifications were rather straightforward; however, the proposal requirements were somewhat daunting. It required providing extensive details on labor, materials, and the overall approach to managing the project.

One of our team members agreed to draft the proposal, while the others worked on the estimating. I had been out of town on another project and returned to the office one day before we had to send the bid to the contracting office in San Diego. On the plane ride back to Vegas, I reviewed the proposal and let me tell you; it was a winner! I arrived at the office feeling great about our chances of success on the project. I immediately met with the rest of the team and told them how awesome the proposal looked. I then asked, did you guys review the proposal specifications to ensure what you did aligns with what the government is expecting? Of course, I got the, "absolutely, yes, sir!"

I started going through the schedule and the estimate and quickly realized my team did not read the specifications as it related to the proposal. Despite our great proposal, the estimate, schedule, and other project information did not align with our written plan. For the next twenty-four hours, all of us scrambled to make the necessary changes to make everything come together. We submitted the bid on time but did not make it past the first round of reviews. Ironically, the contracting officer in our debrief said, "The proposal provided us with an excellent plan for completing the fencing. Unfortunately, the estimate and schedule did not align with your proposal." No matter your spot on the team...read everything!

Reflect: Whenever I think of reflection, it reminds me of my car buying experiences over the years. I go into the dealer, dance with the salesman, sign the preliminary paper saying I am ready to do business, and then wait for him to go and talk to the manager. He comes back, I laugh, the manager comes over, and then I say the words every salesperson in the world dreads to hear, "let me sleep on it, I need to think about it."

What? Think about it! Seriously, whatever you do, please do not leave here without making a purchase. If you do, I know you will realize it is not a good deal, and I will never see you again. Of course, they do not say those words, but we all know it is true. Most of the time, what gets all of us in trouble is not taking a few minutes to stop and think.

You just finished previewing all types of information, raised questions, read a massive amount of material, and now it is time to step back and do some good old-fashioned thinking. Think about what aspects of the documents you did not understand, re-read sections for clarity, or talk to your associates about what you are thinking. Consider different strategies and compare them to previous jobs to see if it might apply to this particular project. Just take some time to think, meditate, or whatever you want to call it. After all, typically, when you stop, and think is when the best information comes to mind.

Recite: Do you ever find it funny the things you remember? While writing this book, I took a cruise with the family. One evening we were out on the deck enjoying some live music when the band started singing some old classic eighties song. Without even thinking about it, I sang every single word of the song, and I have not heard that song in probably twenty years. When the song came out, my friends and I sang it all the time, included it in our conversations, and kept on singing it through high school. I never intended to commit it to memory, and I definitely did not plan on using up storage in my brain to remember the words, but there it was!

The point, if you want to remember something, keep it right on your tongue. Teach it, sing it, talk about it, put music to it, write it down, or continue reading it. Recite, recite, recite! When I am working on a proposal, I write everything down, talk out loud, and converse with my team on different information contained within the specifications. Throughout the years, I have even done little mini presentations on various aspects of the specifications to make myself remember the information. Remember, what I said earlier, your job is to know the specifications better than even those who wrote the specifications.

Review: You previewed, questioned, read, reflected, and recited. It is now time to review! It is time to go back and take a second look at everything to make sure there is a good understanding of the project. The creative juices should be flowing amongst the team members at this juncture. Whom they think would best run the project, or what resources are needed to perform, and even what relationships to leverage to win! One thing I would do is require each of my team members to explain to the rest of the group what aspects of the specifications stuck out to them. Did they read anything they thought is worth sharing? You will know fairly quickly which team members did their homework.

At some point in your conversations with the team, make sure you discuss the budget. I would require them to write down what they think it is going to cost to complete the work. The purpose of doing this is to get an understanding of what everyone is thinking. If one person thinks the job is worth a million and another teammate comes in at ten million, then you might have two very different approaches at work. Make everyone write down any risk, opportunities, threats, they identified in their review of the project. Many times, this little exercise provides a wealth of information.

As a government contractor, you learn quickly; the government thrives on acronyms and expressions. One of my personal favorites is, "let's not ignore the big elephant in the room." In Pensacola, Florida, at the Naval Air Station, I attended a meeting regarding some issues on a project. The project manager that seemed to cause most of the problems was, well, a rather large woman. The meeting began, and after a few minutes, the contracting officer said, "I do not want to waste everyone's time, so I think it best we do not ignore the big elephant in the room." The lady stood up, screamed at the guy, "I cannot believe you would refer to me as a big elephant." After calming down and realizing it is an expression, she apologized for the outburst, and we addressed the project issues. Hopefully, that will help you to remember never to ignore the big elephant in the room when you are discussing critical challenges as it relates to a project. Sometimes we get so excited about winning a project we forget to listen to the voice of reason pointing out a potential pitfall.

If you finish the review process and no one on your team has comments, get a new team. In my entire career there has never been a perfect RFP, specifications, plans, or any other document. There are always issues, discrepancies, or challenges. If your team did a serious review and provides no comments, you need to be worried.

After the review process, there should be lots of questions, discussion, and debate. If not, restart the process because no one has read the documents. Never once in all my career has any agency generated perfect specifications, drawings; it does not happen. If nothing else, someone on the team should identify the key challenges of the project, if not you might be overlooking a material risk. We all get complacent and ignore obvious risk that end up causing serious headaches and, in some cases, costing lots of money.

In 2015, I worked with a contractor that won her first job with the government. I assisted her by bringing in one of the best and brightest estimators I know. Unfortunately, when the team got to the review process, no one asked any questions, or brought up any challenges because it appeared to be an easy project. Each team member went to work and did not communicate. The estimator went to work on bidding all the various aspects of the project except for the metal building. The biggest part of the project, no one even thought about. Needless to say, on bid day this became a huge issue and the team scrambled to come up with a last-minute number. She won the job, but it did not end well based on the inability of the team to identify challenges.

One other point I need to mention here is to make sure you have the team administrator reviewing all of the amendments released during the bid period. There is nothing more frustrating than working on something only to find out the contract changed. Make sure each amendment goes through the same process as the original contract documents.

Always work under the assumption that no one has read any of the documents. I told my team to assume that even the people responsible for the bid did not read what they supplied. Unfortunately, in many cases, we would find that contracting officers or owners did not even know what they included in their specifications. The point…require your team to provide you with their notes from the PQ4R process. Remember, what state we are from, "the Show-Me State."

This meeting occurs hopefully right before or shortly after the job walk. The reason is that most contracts require submission of RFI's within a reasonable period after the job walk. The review part of the process enables you to prepare a comprehensive list of RFI's for the project. After making a list, the team can review to ensure they are legitimate and not answered somewhere in the specifications. If you have no RFI's, you have a problem.

Whenever I talk about the idea of reviewing specifications in detail when speaking most of the people look at me like I am stupid for even insinuating that people do not read the documents. On the surface, it does seem stupid that anyone would place a bid on a project without reading all of the information, but I assure you it happens all the time. If I were to name the number one thing that gets contractors in trouble on projects, I would absolutely tell you it is their inability to read every document related to the bid/proposal.

As a young man, I hooked up with this small contracting company doing a little accounting and other entry level type work for them. The contractor decided he wanted to pursue an RFP, which required a proposal and since I needed work he asked if I would write it. I agreed. I spent hours reading every single document provided and drafted a detail of all the information I needed to address in the proposal. I then went to the owner and his team and attempted to review with them my findings. I thought they would be impressed and interested in my research, not so much. He told me, "Son, I have been doing this for years…just write the proposal and let us deal with the rest. Got it?" I tried to explain, he would not listen. I went ahead and prepared the proposal.

I put together an excellent proposal. I delivered it to the owner the day before the bid and tried to review it with him and his team. He said, "Douglas, thank you for writing this up, here is the check for what we agreed upon. The proposal has nothing to do with the actual construction, so we do not need to go over it." Crushed, I left there thinking there is no way he could be right.

Two months later, I learned his approach did not work. He invited me to attend the debriefing meeting where the contracting officer explained their reasoning for selecting another bidder. The best way I can sum it up is by quoting her, "Sir, when you prepare a proposal/bid/estimate you need to make sure all of your people become familiar with the documents, not just the person writing the proposal."

I could bore you with multiple disaster stories resulting from contractors not doing a thorough review of the documentation. Please, if you only remember one sentence from the entire book, PQ4R the specifications!

Strategy Formulation Step 2: Develop a Work Breakdown Structure

If you followed the procedure outlined previously, then the entire team is engaged and ready to pursue the project. With the team rearing and prepared to go, the next job is to organize all of the team's thoughts into a simple project outline. The outline I refer to is called a WBS or work breakdown structure.

WBS is a scheduling term, and there are varying opinions on its purpose; however, since this is my book, I am going to give you my opinion. First, I think the WBS is not only a scheduling tool but also an excellent resource for beginning the development of a bid/project strategy. According to the King of all scheduling, Mr. Mubarak, the WBS is, "task-oriented, detailed breakdown of activities that organizes, defines, and graphically displays the total work to be accomplished to achieve the final objectives of a project" (Mubarak 2010, 943)." This is the outline everyone on the team should understand and agree on before moving forward. Do not let just one person prepare this and present it to the team make everyone get involved. If you make the entire team participate it will result in significant dialogue, which is great for creating a strategic plan.

I know this might sound a little weird, but I enjoy speaking in public. I love the preparation, the people, the energy, and the fact that something

I say might help just one person. No matter how smart you might be, or your expertise on a particular topic, speaking requires lots of preparation. The best speakers work diligently to organize the material in a manner that keeps the attention of the audience, delivers critical information at the right moment, and works within the time provided by the host.

When I speak, I start with a WBS. A general outline of what I want to present. I go through the framework multiple times, making sure it accomplishes it meets the expectations I set for the presentation. Normally, I will run the outline by a colleague to make sure it makes sense and flows appropriately. Once I establish a solid WBS, then I began to build the rest of my presentation. As you probably already figured out, preparing for a presentation is the same process you follow when establishing the strategy for a project.

For practical purposes, let's build a doghouse. Below is a simple WBS outline for the doghouse.

- I. General Conditions
  - a. Pre-construction submittals
  - b. Meetings
- II. Preparation of Existing Site
  - a. Demolition of trees
  - b. Dirt for site
- III. Underground Utilities
  - a. Electrical
  - b. Plumbing
  - c. Gas
- IV. Concrete
  - a. Sidewalk
  - b. Foundation
- V. Framing
- VI. Roofing
- VII. Exterior Panels

VIII. Interior Construction
  a. Carpet
  b. Paint
IX. Close-Out Procedures

I like my dog. Not the greatest illustration, but it gives you the general idea of what you need to do to prepare the WBS. Again, the entire team should understand the flow. Even people that do not understand the technical aspects of a project should follow the flow prepared at this juncture. If the flow does not work on paper, it will not work on the project.

Every time I present this in front of people, I always have someone ask me if incorporating CSI (construction divisions) is better than a WBS. My answer is still the same; no. CSI does not necessarily represent the flow of every project, and not everyone understands CSI, which means they will nod their heads in agreement to avoid embarrassment, but they will not understand. CSI has a place, and if you are using good scheduling software, you can use this as a filter during the scheduling process.

With an agreed-upon WBS, it is time to start talking strategy with the team. The conversation, led by the bid manager, begins with a simple question, "how do we approach this project in a unique manner that sets up apart from the rest of the bidders?" When you first ask the question, sometimes the room gets quiet because this is not how people typically think when pursuing a project. The bid manager must begin the conversation presenting different ideas or asking questions to make the rest of the team feel comfortable with expressing their opinions. As the bid manager, remember, do not judge anyone for their ideas because if you do, the others will guard what they say to avoid embarrassment. Keep the dialogue flowing by encouraging each team member to express themselves freely.

Let me explain the initial strategy conversation through an illustration most of us understand. After an extended trip or on weekends, my wife, daughter, and I will decide it is easier to go out to eat than to prepare something at home. All three of us agree we need to eat out, but the fun begins when I say, "where would you like to go?" If you are married, you realize this is an awkward conversation that can end up badly if not managed correctly. Tread carefully; do not say where you want to go but ask critical questions to get the wife and daughter talking. Whatever you do, do not discard a recommended location, but rather listen attentively to the ideas presented. Sound familiar, follow the same rules in the strategy discussion. Push, prod, make recommendations, but ultimately let the team decide where you intend on going with the project.

Strategy Formulation Step 4: Brainstorm Competitive Advantages

In my experience, this seemingly effortless practice of collectively thinking is difficult for most owners/managers. I have sat through many of these sessions where the owner/manager seems inconvenienced, which is sad because that sends a message to the team, "he/she does not care what any of us think." Fully engage in the brainstorming session and include every team member. Remember the reason for the brainstorming to develop the most effective bid strategy, which will provide a competitive advantage against the other bidders. The team is trying to figure out how to correctly apply the strengths/weaknesses of the organization to this specific project.

In my current position as a forensic accountant, I work primarily with sureties that have a contractor (Principal) either in default or experiencing difficulties on a project. When I first arrive at their office to perform a mini-audit of the projects, I first meet with the Principal to get an understanding of their thoughts in regards to the issues with the project(s) in question. It always goes about the same, the Principal spends the majority of our time together blaming anyone and everyone involved with the project, ironically, none of them ever accept responsibility.

After a barrage of excuses, the Principal usually tells me they have a meeting to attend and hand me over to their staff for further communications. When the door shuts, and the Principal is gone, the real fun begins. Without exception, the staff starts telling me all kinds of details the Principal conveniently forgot to mention. In detail, they explain why the project failed, what they need to do to resolve the issues, and sometimes tell me what the Principal is doing with the money!

I always wonder why these people do not tell the owner all of the things they share with me? I mean, the guy/gal pays them to assist with managing the company, and they watch as the company suffers and say absolutely nothing. Honestly, I know the reason. It is the culture. In the past, the owner/Principal disregards their comments, ignores them, or lets them know through his/her actions he is not interested in what they have to say. Eventually, they accept the fact their opinion holds no value, and they quit talking. If you are the owner/Principal/contractor reading this, pay attention, hire smart people, and listen to them!

Let me define brainstorming. It is an unpredictable, unplanned storm in the brain with lots of surprises! When I think of brainstorming, I think of the hurricanes I endured while living in the Panhandle of Florida. For days, the weather stations give you all kinds of predictions, where the storm is going to land, the wind speed, the potential water damage, when it will arrive, and how long it is going to stay. Most of the time, they end up getting it all wrong, because the truth is no one really knows what might happen with a storm, which is why it is called a storm!

To me, brainstorming is one of the most essential tools a business has at its disposal, yet it is one of the most neglected tools in most operations. Owning the company, having the right last name, or some fancy title does not mean you have all the answers. If nothing else, remember every single person approaches challenges with a different perspective, which in many cases, is what helps make the right decision. I digress.

A good brainstorming session follows the model of any storm. If you live on the East Coast, you know how nice the weather gets right before the storm, the temperature drops, the wind blows, the clouds roll in, and for a time it is very peaceful. It is referred to as the calm before the storm. This is the environment you need to create before a brainstorming session, the team needs to be relaxed, focused, and in good spirits.

A good brainstorming session follows the model of any storm. If you live on the East Coast, you know how nice the weather gets right before the storm, the temperature drops, the wind blows, the clouds roll in, and for a time, it is very peaceful. It is referred to as the calm before the storm. This is the environment you need to create before a brainstorming session, the team needs to be relaxed, focused, and in good spirits. In one of my consulting arrangements, I visited an office to observe a brainstorming session. No one in the meeting said anything, which seemed weird. After the meeting, I found out the owner just read the riot act to them and then told them to meet in the conference room for a brainstorming session. It does not work like without the calm first. When the team is relaxed, the session begins.

Storms do not work on timelines. I remember one of my new contractors called me and told me he tried brainstorming, and it just did not work for his team. I knew a couple of his people, so I did a little backdoor inquiring. He called all of his people into the conference room, told them he had about twenty minutes, and then at the end of twenty minutes, he closed the meeting. Yes, managing time is important, but if you want the results, you must not restrict the time.

In 2005, hurricane Danny nailed the Gulf Coast. Unlike other storms, it came through right in the middle of the day giving everyone a front seat to its powerful winds and rain. I remember watching from my living room as it ripped my fence apart, tore off my awning, and blew a tree over into my pool. As all these things happened, I never once went outside and screamed at the storm, letting it know my thoughts on its damage to my personal property. In a storm, unpredictability is anticipated, and no matter what happens, no one questions nature. In the brainstorming session, let people talk, no matter how stupid what they say might be, let the dialogue flow! Just observe, do not question anything, or shut anyone down for anything they say. No matter what they say, write it down, and act like it is genius! If you do this, your people feel comfortable, and they will give you their best.

I realize the world is crazy, but I do not think anyone has sued Mother Nature recently. The BP oil spill that occurred in the Panhandle a few years back ended up costing about $61 billion, and BP nearly collapsed over its involvement in the catastrophe. Businesses and people retained attorneys and filed claims for all types of damages from this spill. Ironically, Hurricane Katrina caused $81 billion in damage, and Mother Nature did not even receive any bad publicity on the matter. They did not even cancel Earth Day! When conducting the session, it is a storm—and thus there are no repercussions for good, bad, or stupid ideas. The point is to only record the events, or in this case the ideas. Do not let anyone attack an idea, a person, or anything for that matter. Also, make sure no one in the room intimidates the team.

For all of you organization experts, don't worry, the storm calms down, and when it does, the weather is perfect. One thing I always noted about hurricanes is the great weather after the storm passes. In 2004, after Hurricane Ivan devasted the Gulf Coast, we had the most beautiful two weeks of weather imaginable. Unfortunately, you cannot enjoy it because of the damage, but it is still great weather. After a brainstorming session, this is how everyone should feel. If people are frustrated, then you did something wrong. Remember, the purpose of brainstorming is to get ideas, not to debate them, or question them, get ideas!

Just a few more comments on this session. Before starting, make sure you explain the rules of the session. If you would like, use the storm analogy. Most importantly, make sure every participant understands their role and the rules of engagement. During the meeting, write everything that is said on a big board. Ask questions to spark thinking, but only lead the conversation to get the meeting off the ground. Make sure there are refreshments and absolutely no distractions during this time frame.

Strategy Formulation Step 3: Establish Project Goals

Whenever I present this section publicly, someone always asks me, "Why do you brainstorm before you set goals?" Great question. If you establish the goals before the brainstorming, the team will only think in the direction of the goals. A lot of good ideas are overlooked because owners/managers say too much too early.

Back to goals, establishing project goals is crucial to developing strategy. Think about it, how can you create a plan to accomplish something if you have no idea what you want to achieve. Whenever I consulted with a company working on an estimate, I would always ask, "what is your goal on this project?"

In the previous step, the entire team participated in identifying potential competitive advantages specific to the project you are pursuing. As you establish the goals, consider the ideas presented in the brainstorming session and see how you can leverage those advantages in obtaining the project objectives.

Set goals for every aspect of the project. What is the margin you want? What are your self-performance objectives? Does this project give you access to a new client or allow you to obtain some past performance you previously did not have? There have to be specific objectives related to the project, so you can develop strategies to reach the goals. Planning without goals is like asking for directions when you have no destination.

When I owned my construction company in Nevada, I worked with an exceptional group of people. Most of what I write about in this book I learned while working with each of them. I thoroughly enjoyed watching each of them establish specific goals for projects. The estimator would develop buy out targets, the superintendent tried to figure out methods to accelerate the schedule, and my administration team always would openly discuss efficiencies. All of them understood the competitive advantage of the project from brainstorming, and they incorporated it into figuring out what they wanted to accomplish on the project.

There are lots of studies out there supporting the idea of establishing goals in a group setting. Common sense says if a group of people establish a goal there is automatic accountability, meaning the chance of reaching the goal is much higher. One thing for sure, it is really cool when the team establish project specific goals and at the end of the project, they reach them!!! Great stuff. I could go on for hours on the importance of setting goals, I challenge you to do some research.

Strategy Formulation Step 4: Assign Duties

It goes without saying, everything going on thus far should be documented. If you remember, the project has an administrative person assigned and they should be keeping track of everything that has occurred. All of this happens very quickly, normally within a few days, so it is imperative the administrator keep up with everything.

At this juncture, the bid manager should have a good understanding of his/her team. The WBS meeting, the brainstorming, and the goal-setting provide insight into what each team member does best.

It is now time to get to work. The bid manager needs to assign his/her expectations to each team member to ensure they understand what they are supposed to do and when it should be complete. In my company, we used the WWW (Who? What? When?) Sheet to make sure everyone remained accountable.

Who?	**What?**	**When?**
Ted Phillips	Complete Schedule	1400 hours, 5 September 2017
Lisa Frank	Complete bid bond application	0900 31 August 2017
Ted Phillips	Do self-performance take off for fencing	1400 hours, 7 September 2017

This single document is one of the essential components in bid preparation. I could write a separate book on all the times I have been a part of bid day where someone forgot to do something, and the person in charge forgot to make sure it did get done. I always kept the WWW? posted in plain sight for the whole team to see. During every project meeting, this is the first document reviewed.

As a general recommendation, I would require my bid manager to review the WWW every morning before starting the day. If the team does not take this seriously, then you need to get a different team. Bidding is hard;

two weeks to compile an entire bid/proposal, which might make the company great or might be the last project ever done! Take the process seriously!

A good bid manager keeps the WWW updated and ensures the items on the schedule get completed timely. If the current bid process is stressful, then put this tool into practice and make bid day a little more bearable.

Strategy Formulation Step 5: Evaluate Strategic Plans

Brainstorming done, goals set, it is time to bring it all together. How do you plan to win this bid, what are your project goals, and how will you get there? The entire team should be fired up at this point, excited about the prospect, and ready to apply all of what they have individually worked on thus far. If you get to this point and there is no energy, you might want to head back to the go/no go section and reevaluate.

Thinking, as described in the paragraph above, is a little different than the average contractor thinks in regard to bidding. Many contractors believe I need to make sure my bid is lower than everyone else. That philosophy is scary because it means the product delivered is a commodity, and we all know how a pricing war ends.

We just finished going through a rigorous process to make sure we had a high probability of winning this bid – so the question is how do we win? What is the plan? The key, as discussed previously, to increasing the "kill rate" (acquired versus pursued) is by going after those projects where we can leverage our competitive advantage to get the win.

The bid strategy outlines the unique approach to successfully delivering the project, which spills over into how your bid is going to look better than all the other submissions. Think about it this way let's say you pursue a fencing project. It is a Government project, LPTA (lowest price technically acceptable) proposal. If you want to win, then the bid must come in lower than everyone else, right? So, do you just cut, cut, cut until there is no margin left? Of course not, this is where you try and figure out how you can do the job uniquely or different than the other bidders. If you get in the habit of just reducing your numbers or margins, you will slowly remove yourself from the market. What is it you can do better than anyone else, which can act as your competitive advantage in the bid process?

Remember, earlier, all those exercises on identifying strengths, weaknesses, opportunities, and threats? Guess what? It is time to put those concepts to work in figuring out your bid strategy. Let's go back to the fencing illustration. For illustration purposes, pretend one of your strengths is capital. Meaning, the company is doing great financially. With that strength, maybe you approach the fencing supplier and work out special terms because the firm is capable of paying earlier than the other bidders. Now, you provided yourself a little edge over your competitors based on entirely on internal strength.

I love doing demolition; it is by far my favorite part of contracting. In 2015, the local Air Force Base released a demolition project for three separate buildings, two small buildings, and one sizeable four-story building. When they released the bid, my team and I walked through this entire process and decided to pursue the project. Our competitors, through word of mouth, were mocking us because going after a demolition job, competing directly against subcontractors that do nothing, but demolition seemed like a waste of time.

Of course, my team and I went through this project and developed a strategy based on two things: an opportunity, and a strength. First, let's talk about the opportunity. About two years before the release of this project, my project manager made a connection with an independent contractor that owned his excavator and did nothing but excavator type work. He had no interest in hiring people, growing his company, or dealing with any aspect of a business. He just wanted to run the excavator. For two solid years, his name came up every time we evaluated current opportunities. On this day, we leveraged our opportunity and met with this guy about doing the work. A key strength of our company involved managing government contracting, so together, we had a great plan.

A second strength we listed on every single project is our innovative ability to think outside the box. The team I worked with did some crazy brainstorming, and through the years, we came up with some very unique approaches to winning projects. With our excavator operator in place, there remained one hurdle to competing with subcontractors that did demolition exclusively - hauling. Even with the operator's competitive advantage, we would have to hire our competitors to haul. Of course, they would increase their hauling prices, which would eliminate us from the competition.

In our brainstorming session, someone said it would be cool if the dump trucks could park on-site, we fill the trucks up, and then they dump. The excavator would never be waiting on the drivers, and we would eliminate paying drivers to go back and forth all day long. I think she might have been joking, but I took the idea to heart and came up with an innovative way to perform the demolition efficiently and save money on fencing the entire project.

Hello, my name is Douglas Allen and I would like to speak to your owner about a business proposition. The lady on the other line said, "Sir, he is not here right now, but I will have him call you shortly." This is the message I left every container company in town. Two of them called me back. I explained I would like to rent every dumpster/container they had and line the entire outside of the project with them, which would secure the site and allow my excavator to fill them all day long. The dump company could come at 3 pm every day and empty as many of them as they could until the dump closed. We came to terms and eliminated the hauling.

Not only did we win the project, but we also beat all the demolition subcontractors by a fairly decent spread. One of the subcontractors filed a protest with the government saying we underbid the job and would not perform. We performed the job, made a phenomenal margin, and enjoyed every minute of gloating to our competitors.

Push your team to approach every bid uniquely. Focus on applying strengths and opportunities in every job you pursue. Most contractors focus on figuring out their cost, marking it up, and submitting the bid.

They figure if they do this enough, they will eventually win. Even a blind squirrel gets a nut once in a while, right?

The last contractor I assisted met with me and told me developing a strategy changed her approach to bidding. She was so excited to show me what she had done and invited me to review the bid with her team. As a small contractor, she competed with just about every new and emerging business out there, which meant margins were tight, and the jobs were not very big. Strategy became very important.

On this particular project it demanded a lot of material. In her brainstorming session, she along with her team decided if they funded all of the material it would eliminate the markup from all the subcontractors, which would provide them a competitive advantage and hopefully a win! They priced out all of the materials, and requested the subcontractors only bid the labor. Sounded pretty solid, and by looking at the pricing it looked like they would win.

After about two days I finally sit down with the owner to discuss her strategy. I walked through her process and asked a few basis questions. I then asked if I could review her financials because the strategy, she intended on pursuing required a solid balance sheet. She fumbled around and printed me a balance sheet from her accounting software. I took one quick look and realized she had a problem; the balance sheet would not support her strategic plan on this project. Bottom line, her actual capabilities would not enable her to pursue the selected strategy.

She developed a great strategy, but she overlooked the analysis part of the process where you identify strengths, weaknesses, opportunities, and threats. After I explained the requirements for financing a project, her face told me she had no idea as to what this entailed. A strategy must be based on reality. Conceptual plans are great, but they must be realistic. Remember, we function in the "Show Me State," meaning test your concepts and make sure they work. I have seen many contractors lose their business because they get wrapped up in confirmation bias, where they accept inaccurate evidence to support their unrealistic plans.

What I am about to talk about might seem a little obvious as you read; however, I run into this all the time, so it warrants some attention. Every year I attend an event at my alma mater in Florida. The three-day event always lands on the same day a bid is due. For several months I had worked with a young lady in my office training her to manage the bid process. After several test runs, I felt like the time had come for her to take the reins and manage a bid in my absence. I would be gone on bid day, so I figured I would let her run the entire process, and I would check in periodically.

Every day she would call and give me the low down. When she got to the strategy part of the process, she called to explain to me the plan her team had prepared to win the project. The strategy included:

1. Obtain the best numbers from subcontractors
2. Complete the work efficiently.
3. Create an aggressive project schedule.

I must say, sometimes managing is frustrating. In my head, I thought after all the months of training and discussing strategies, and she gives me three things every company should do on every single bid no matter what! I kindly asked her, "Do you think these three things are going to set us apart from our competitors, provide a unique advantage?" The other end of the phone line, silence.

I must say, sometimes managing is frustrating. In my head, I thought after all the months of training and discussing strategies, and she gives me three things every company should do on every single bid no matter what! I kindly asked her, "Do you think these three things are going to set us apart from our competitors, provide a unique advantage?" The other end of the phone line, silence.

Do not mix standard procedures with strategy. It is easy to think your operation does things differently than everyone else when, in reality, all of you are just executing best practices. Best practices or standard operating procedures do not provide a competitive advantage in most cases. Remember, with no strategy, you are nothing more than a commodity competing with everyone to see who bids themselves out of business first.

Strategy Formulation Step 6: Identify at Least Three Strategic Plans Providing Competitive Advantage

On most projects, the team comes up with three to five great ideas for bidding the job with a competitive advantage. Unfortunately, many companies also do not keep these ideas at the front of their minds as they work through preparing the bid/proposal. If the strategies are going to work, then you need to bring them up in every conversation, scheduling meetings, working with subcontractors, evaluating the proposal, and in all of your daily encounters. Make sure the strategy is in front of everyone all the time. The worse thing to do is have one team member focused on one plan, while another team member is doing something completely different.

As mentioned previously and addressed later in another chapter, the bid manager should conduct daily briefings reviewing the Who? What? When? schedule and talking about the project strategies. Again, the objective is to make the team implement the strategic planning into the various assignments.

As you read, you may think there is a lot of time consumed in preparing a bid strategy. Despite how it is written, the process is relatively quick if the team spent the necessary time early on gaining an understanding of the project. The return on preparing a strategy is phenomenal because as your kill rate increases, you become in control of the process, which frees up a lot of time and energy.

General Strategies

For me, speaking about strategy without introducing the Grand Master of Strategy seems foolish. I have never met Mr. Michael Porter, but I have read just about everything he has ever written on strategy and strategic planning for businesses. I would challenge you to do your own research on strategic planning, starting with his books.

In his book *Competitive Advantage*, Mr. Porter introduces three generic strategies, "...three generic strategies for achieving above-average performance in an industry...cost leadership, differentiation, and focus"

(Porter, 1985, p. 634). I want to discuss these three concepts briefly to assist as you work on developing your strategies specific to projects.

Cost Leadership

As an accountant, it is frustrating to observe people misuse terms. For example, cost and price. If you go out to the store and purchase an item, the price is what you paid for the item. If you buy something to resell, then what you paid is your cost, "[Cost is] economic value of resources consumed in making a product, process, or completing an activity, [whereas price] is what a seller charges a customer" (Merchant 2017, 93).

As an accountant, it irritates me to hear people loosely use the words *price* and *cost.* It is even more frustrating when speaking with estimators or bid managers who use these terms as if they are interchangeable. Price and cost are two separate concepts and must be treated accordingly. Cost represents "the economic value of resources consumed in making a product, process, or completing an activity, [whereas price] is what a seller charges a customer" (Merchant 2017, 93).

When presenting in person, contractors tend to get frustrated when I introduce the "cost leadership" strategy. They reference my earlier comments about becoming a commodity and consider this strategy the road to going broke. Most of the time, they think this because of the miscommunication with the words cost and price. The strategy is cost leadership, not price leadership. The strategy deals with operating efficiently, or working with economies of scale, or providing services for a lesser cost due to value engineering. The cost of doing the job is lower than competitors, meaning the company will obtain a more considerable margin based on this competitive advantage.

On a lot of smaller jobs, a low-cost strategy is the only option available. For example, if you pursue a $200,000 RFP (request for proposal) that includes a standard scope of work, it is challenging to attempt to differentiate the company. What is important to note regarding the low-cost strategy is this is not a one time deal, meaning you do not select this strategy based on one project. Pursuing this strategy requires a commitment to keep the company cost low, operating efficiently, with little overhead.

In 2010, the company I worked for had an excessive amount of backlog, a good problem. I placed a few advertisements for project managers and received just over two-hundred responses. After filtering all of the applications, I narrowed my search down to ten people. All ten of the candidates possessed considerable experience in construction, and most of them had either owned or been a shareholder of a previous company.

As I interviewed them, I began to see a common theme. Every single candidate had worked/owned for a large entity, which tanked quickly when the market turned. As I spoke with them, I asked what happened to bring down their operations so quickly. Each one without hesitation said we had too much overhead and could not operate without maintaining a large backlog of jobs.

In construction, it is imperative to keep an eye on overhead. No matter how big you grow, it is essential to understand all of the expenses associated with running the business and managing projects. Companies surviving difficult markets understand the importance of managing overhead and quickly adjust the model based on operations. Pay attention to cost!

Working with estimators on establishing actual cost is sometimes a challenge. The estimators working for me over the years would sometimes base their calculations on previous experiences, guesses, and industry standards. When the basis of your strategy is cost, that is what you need, cost! Do not let anyone add contingencies to line items in your bid unless the entire team reviews and agrees. I could tell you several stories about jobs I lost by small margins only to find out that an estimator included some contingency because he/she did not feel like doing the research to identify actual cost. No contingencies, unless the whole team agrees!

In the bid process includes subcontractor invoices, make sure you scrutinize the numbers. Many times, subcontractors add contingencies or make assumptions, which increases cost. Review the estimate with them in detail and find out what they added or forgot to add in some cases. If they change their proposal, request they resubmit it to you with the changes. Amazingly, subcontractors are notorious for not recollecting changes related to dollars.

If you intend to pursue a low-cost strategy on a project, there can be no assumptions. When my firm first graduated from the 8a program, we had no choice but to pursue a low-cost strategy, so I hired a well-respected estimator and explained to him how we intended to win through the use of a low-cost strategy. Initially, he worked very closely with me and provided accurate costing information based on documented sources. After a few bids, I made the mistake of not checking his work, and it cost me dearly. We pursued three separate projects in Bakersfield, CA, at a remote military base. We lost all three jobs by pretty good margins. I requested we do an autopsy of the cost proposals to identify why we were high. The analysis indicated we were very high on general conditions. A little further research, identified the estimator plugged in numbers from a job we bid on Los Angeles, not verifying based on the location of the project. If my estimator had verified those costs, we would have been awarded all three projects.

One argument I seem to engage in quite frequently relates to my position on contingencies. Before you get angry and start telling me there is no way to bid a job without contingencies because of the unknowns, let me explain my reasoning.

First, I do not like contingencies because, in my experience, they are sometimes nothing more than an excuse to avoid doing the work to determine the actual cost. For example, I remember working with an estimator on a large sodding project. He included a contingency on all labor for small tools, which significantly increased the labor cost. I have thrown my share of sod, and I know you need rakes and a strong back. Based on the contingency he included, I could have bought brand new rakes every morning they worked and came out cheaper. To me, that is just lazy.

My second objection to contingencies is how they get added to the estimate. I do not want anyone making decisions regarding my money without involving me in the decision. When estimators include random contingencies, it becomes difficult to understand the bid and make the right decisions on your strategic planning. Please do not add anything to my proposal, unless you explain to me in detail your reasoning. Many jobs are lost/won based solely on an estimator's perception of contingencies.

NEVER, NEVER, let an estimator or anyone for that matter make decisions for your company. If contingencies are necessary, that is the owner's decision to make!

Final comments on implementing a low cost strategy:

1. Manage and monitor overhead. Make sure you keep your numbers in line with industry standards or lower. Try to operate with variable expenses, not fixed.
2. Only deal with accurate numbers – no assumptions, no contingencies, no guesses.
3. Evaluate every line item of the schedule ensuring all parties provided real numbers based on the conditions specfic to the project pursued.
4. Don't forget where you live, "the show me state". Review everything and require documentation. In God we trust, everybody else bring documents.
5. At minimum, twice a year evaluate cost on projects. Ask yourself, does this cost add value to the project, and/or would my client notice if the cost no longer existed.
6. Never allow anyone to calculate general conditions on a percentage format. Always review in detail the direct/indirect cost specific to a project.

Low cost is an excellent strategy for contractors if they know how to manage money. Just like every strategy, it requires thought, planning, and execution. You don't decide to pursue this strategy overnight; you must plan and set your operations up accordingly.

Differentiation

The second generic strategy from Mr. Porter is differentiation. Executing a strategy of this nature requires the company to set itself apart from competition based on its ability to perform, unlike its competitors. Sometimes this might be specific to one job, or in other cases, it might apply to every project completed.

For reasons unknown, my company got into building public park restrooms. The local municipalities kept releasing bathrooms projects, and we kept winning them. The more we did; the more competition would get closer to duplicating our numbers. Quite frankly, the low-cost strategy won us the first few of them, but it did not take long for the competition to make adjustments. If we wanted to keep doing them, we needed to do something differently, or we would have to keep reducing our price, which is not the ultimate goal.

In a brainstorming session, one of my superintendents came up with a great idea. Instead of building the bathroom, why not get it pre-manufactured and dropped on site. We sent off the specifications to a manufacturer to see if it would work, and they said, "no problem." We won four more-bathroom projects based on doing something completely different. After the fourth project, the competition caught on, and it became a pricing war - we moved on to other projects.

When it comes to differentiation, the key is to try and do something different than everyone else, unique! My company leveraged tools for planning projects. Every project required a schedule as part of the bid process because I believe there are two crucial elements for project success: time and money. If you manage these two, the likelihood of project success is high. We bid a job in Wisconsin, and in the RFP, the owner allowed contractors to split the job in two, half of it done before winter, and the balance complete after winter. It seemed logical, and I also knew that everybody would bid it accordingly. Not us, we needed to do something different to win, and we did. We put together an aggressive schedule, which included some additional crews and some overtime, and had us finishing before winter. Not only did we win the project, but the owner loved our work and requested us to perform several other jobs for them directly.

One final comment on differentiation. Sometimes companies perform in a way that makes them different or unique all of the time. It might be a patented process, or technology, or proprietary processes or products. If this is something your company offers, then focus on building your business around whatever makes you different. Be careful; make sure the product/service, which is different, is sustainable because I assure you someone out there is working to duplicate what you created.

Most of the projects my company bid incorporated some level of differentiation. As a general contractor, we always focused on a merger of both cost leadership and differentiation. As part of the decision process in deciding to pursue a project, I tried to keep these two strategies in mind, and it paid off. We won all types of projects throughout the country because we bid strategically.

Focus

The final generic strategy deals with differentiation through focus. Focused differentiation deals with providing a unique service to, in some cases, a select market segment. A good example, contractors that only do restoration work on historic buildings. It requires an intense focus on performing only one aspect of construction. Contractors struggle with incorporating this type of strategy because it requires disciplined attention to doing only one task generally for a small market.

Okay, please sit down, I am about to reveal another crazy Douglas Allen story. After my first college degree, I had a crazy dream I wanted to be a mechanic. Yep, I became intrigued with understanding how things worked and repairing them. It started as a hobby, but it led to me joining a dealership and working as an automotive technician. After my grunt work period, I obtained all my certifications and held the distinguished title of Master Technician and became the shop foreman of a large dealership. In this role, I assisted the other technicians and worked on cars myself.

The automotive industry is unique when it comes to compensation. Every job pays a specific time, and to make real money, you must beat the time allocated to perform the task. I observed all my peers and quickly figured out how to do very well with my hours.

If you have owned a car in the last ten years, then you are familiar with recalls. The manufacturer sends you a letter, tells you to take your vehicle to the dealer for a no-cost repair. On the other side of the house, the dealer gets notification providing instructions on how to complete the repair and how much time they intend on paying the mechanic to perform the repair. The time allocated for warranty and recall work is significantly less than when the customer pays for repairs. Most technicians hate recalls and dread doing them.

Most of the time, technicians would get so angry about doing recalls they would take twice the time to complete the work, which made the situation even worse. Watching the frustration, I got an idea. I want to the Service Manager and the dispatcher and advised them to give me all the recalls moving forward. They agreed. Everyone thought I had lost my mind, but I had a plan.

As soon as the manufacturer released a recall, I got the paperwork and reviewed the procedures required for the repair. I built three square wooden boxes and placed the appropriate tools in the box needed to complete the repair. The first few recalls performed, I timed myself to ensure each time I improved a little bit. After about the third or fourth recall, I knew the exact time it would take, and then I would knock the recalls out all day and do very well! For three months straight, my production soared above the rest of the technicians, even with my duties as foreman.

I focused on one task, making me become an expert and increasing my efficiency to a point it became a competitive advantage. Sometimes in contracting, this approach serves as a great approach to the market. It sets you apart and enables you to charge a premium because you are an absolute expert at what you do. No one else can compete with you because of your economies of scale and efficient operations.

The quick strategies presented here, low-cost, differentiation, and focused differentiation, give you a starting point. The most important thing you can do is figure out what or how you intend to be different from your competition on each project.

CHAPTER 4

CREATE THE PROJECT SCHEDULE

Thus far, a majority of the concepts presented are already part of what you do, whether you realize it or not. Many times, the steps outlined in the first part of the book occur, just not formally. After reading the book, my hope is you will formalize your bid process by incorporating the methods discussed. Throughout the book, I have referenced the three-legged stool to simplify the critical success factors of a construction company. All of the legs, getting the work, doing the job, and keeping score, are equally important; however, the consequence of not getting the work is two-fold. If you do not bid correctly, you never win a job; if you bid wrong, you may lose your opportunity to bid again. Pretty risky!

The next part of this book is where I kind of change gears and present a new concept as it relates to preparing a bid/proposal. When I do public presentations on my book, this is the section where people tend to agree with the concept, but then disagree with how I recommend proceeding. Let me give you the scenario as it occurs in the presentation setting.

First, I go through everything already discussed, and people are writing notes and thinking of ways to apply what I tell them. Then I start talking about the two most critical aspects of a successful project: time and money. I give illustration after illustration, and the heads are shaking like crazy in agreement. I proceed to talk about understanding costs and determining all of the expenses associated with the project, including direct and indirect. Then I jump on the topic of time and explain how important it is to understand the relationship between time and money. Still, lots of smiles, head nodding, and body language saying, "preach it brother."

Everyone in the room is excited, feeling great about what I just told them, and then I go and ruin the mood by saying, "Why is it we do not prepare schedules in the bid process if time is one of the two most critical factors on the success of a project?" The look in the room is bewilderment, kind of like, I agree, but wouldn't that require a lot more work. Yea, of course, but as we discussed, we want to improve our kill rate, only pursuing projects that we likely will win, right?

Remember earlier, we discussed the two philosophies: (1) bid like crazy, and statistically, you will win enough jobs to stay busy, or (2) a bid based on a calculated process pursuing only what you know you can win. It is decision time; how do you want to run your company? Do you wish to employ a group of professionals and have them pursue work with the hopes that statistics fall within your favor winning hopefully 10%, or would you prefer to staff your team with those focused on results and winning more like 70% of what they pursue? Most likely, I have your attention now as well. Lets talk through this in detail.

If you were to speak with an estimator about scheduling, they would most likely tell you to talk with either the superintendent or the project manager. They might even tell you the schedule is something prepared after the owner awards the contract. Ironically, they are developing a cost estimate for work-based primarily on time as the cost driver. What do I mean? In management accounting, we search for either resources or activities that drive costs. For example, if labor is your cost driver, then every hour worked increases costs.

With that said, in construction, there are multiple cost drivers mainly related to time. Also, the contract stipulates a time for performance, which is a constraint. With just those two concepts in mind, how does an estimator prepare a bid without consideration of time? Let me explain. What they do is break out various activities, determine a cost for the operation, and then sum all of the results. Most of the time, they make assumptions about the time, figuring the owner provided ample time to complete the project, or they will add the activities together and throw a good guess at how long it will take to complete the work.

Before I go any further, let me go ahead and entertain what you might be thinking. If you are an estimator, you probably thinking, "that is absolutely not true, we consider time in all of our calculations." You are right, you consider cost as it relates to each individual activity, but construction is not about just one activity, but instead merging thousands of activities together with the hopes of delivering a project. Also, if estimating includes obtaining a number and then backing into a time calculation, how does the estimator know if the price is accurate. Let me explain.

When I first started working in construction, I observed the process firsthand with a pretty large contractor. First, compile all the subcontractor's proposals, choose the best based on price, reputation, and exclusions/inclusions, then add on general conditions and margin and submit to the owner and hope for the best.

After observing this practice, I had a unique opportunity to audit some proposals to determine the variance between the estimate and the actual costs. One observation I discovered multiple times is the incorrect calculation of labor. For example, a project might include a six-month time frame, and the subcontractor labor cost would cover almost the entire period despite they only worked for a few weeks. No one calculated the dollars against the time, meaning the subcontractor got paid a lot more than what he earned. If someone had checked this against a schedule, they would have identified the overage and requested the subcontractor make the necessary adjustments.

Most of the time, estimators primarily focus on the dollars required to complete the project. Breaking the project down into activities or line items and then determining the cost seems like an archaic process taking way too much time. If an entire business is dependent on either winning projects to stay in business or making sure the win does not put you out of business, then wouldn't it make sense to invest the extra time. What I find ironic is the argument related to time? Think about it, if you have two weeks to submit the bid does the time it takes to figure out the cost change? No, you will spend a solid two weeks working on the bid whether you focus primarily on the money or if you implement what I am proposing and involve time in the estimate process. Either way, you will expend the same resources. Not to mention, consideration of time and money provides a competitive advantage over those only focused on money.

I am an accountant. My student loans illustrate my commitment to the topic. I think like an accountant, a lot of what you read from here on out sounds like it is for accounting, but if you will apply what I am presenting I assure you the bidding process will become more efficient, your company will win more bids, and you will make more money and sleep a lot better if you are the owner.

Early in my consulting career, a construction firm retained me to assist with establishing a cost management system. After a few months, and radical changes, they decided it would be a good idea to hire me to oversee operations. They hired me and gave me a title, "Program Manager". To this day, I have no idea what the title means, but I figured I would continue to work on internally processes and work under whatever title they selected. Previously, I established a solid cost management system, so now I needed to take a good look at operations.

As with any assignment, the first thing I like to do is take an inventory of the existing processes. I want to understand how things work currently, and get an idea of what is broke, where the bottlenecks are in the process or find out if the company even has a known method of performing. As an employee, at this juncture, the most effective tool became the interview. I met with all the department heads, project managers, and superintendents and asked questions.

As in most construction companies, the interviews ended up with one department blaming the other and vice versa. Earlier, we talked about the three legs of a construction company: getting the work, doing the work, and keeping score. The people doing the work blamed the estimators for all the site problems, and the estimators blamed the site personnel for not executing the plan, and the office personnel thought the estimators and the field personnel were total idiots. A typical day in the office of a construction company, right?

The estimators and the office personnel communicated their positions eloquently, so I decided to address field issues first. I visited every single project in play at the time and met with all of the field personnel to review the operations side of the house. Quite frankly, after observing the miraculous work of the field to maintain the schedules and budgets of each project, I knew I had missed something. On my last job visit, I met an older gentleman that worked for us as a superintendent. After a few minutes of conversing, he explained he had just two weeks left of work as he planned on retiring at the end of the month.

I guess because he planned on leaving, I felt comfortable unloading on the poor guy. I told him everything about the interviews, all the complaints, and my frustrations with trying to figure out how-to advice on managing the operation. He graciously listened to everything I told him, and then said, "Mister, I feel for you. I would not want to do your job. I do not know a lot about business, but with forty years of construction, I will give you one bit of advice, forget everything you know when it comes to construction and focus on two things: time and money. You do these two well, you will be successful." He stood up and directed me with his arm to the exit.

Wow, I got back to the hotel, and it felt like someone just secretly gave me a cure to some significant disease. Projects succeed/fail based primarily on time and money. Think about it, if you perform technically, I mean you have the best quality known to man, does that mean the project is a success if it runs three months over on time, and $500,000 over on budget? No, of course not. All of the documents, plans, specifications, etc., we promote as being the most crucial aspect of a project do not necessarily determine success. It comes down to managing all those items through the most essential success factors of a project, time, and money.

Great, I understood the formula for consistent success, but how would I apply it to a construction operation. Obviously, based on my communications and the general consensus of the industry there is a disconnect between what happens in the office versus the field, how would I resolve the disconnect. Think about it, there is already daily reporting, meetings, quality control procedures, and people that knew much more about construction and projects then I would ever know.

I figured my best move at this point revolved around researching some larger successful operations. The firm I worked with operated as an 8a, which gave me opportunities to interview and speak with some of the largest firms in the country. Whatever I needed; these companies would provide because they hoped we would partner with them to pursue government projects. After a few months, we entered into a joint venture with a large successful contractor and I knew it would only be a matter of time before I figured out how this large contractor successfully carried out operations. What I found is not exactly what I expected.

In a formal mentor/protégé relationship the larger contractor is required to take the smaller contractor under his/her wing and teach them how to manage a large operation. Honestly, our mentor did a great job and opened operations for my team to evaluate. We started by observing the office practices, including the acquisition strategies, bidding, proposal writing, subcontract management, and oversight of field operations. I met some incredible bright people that impress me to this day. The office team for this company possessed some of the smartest construction professionals I had ever met. Ironically, their largest complaint related to the inability of their field people to effectively manage and run projects. They complained about the difficulty in finding professional field personnel to oversee, manage, and deliver quality/safe projects. The people they employed did an okay job, but the office definitely felt as if they carried the torch for the company. After about six weeks in the office, I tended to agree. The mentor decided to send us out on projects to observe field operations.

Even after all my time in the office with the contractor, I still did not identify the missing link between the departments, nor had I determined how to create congruency between the field and office. After spending the last few months in the office, I started to think maybe this is just a characteristic of the industry. I flew to Arizona to spend three months on a project, and boy did the learning continue.

I hit the ground running, attending site meetings, reviewing schedules, and meeting all the people I assumed were idiots! These people were amazing, they made things happen despite all the negativity from the office. In some cases, I watched them completely modify directions from the office, or change the schedule on the fly without even doing any analysis. They understood the subcontractors making them perform, they carried out a different schedule than what the office reviewed, and in many cases did or authorized activities with little or no consideration from office personnel. As with the office personnel, they continually blamed the office for the field issues and called them paper pushers, white collar idiots, and disliked them equally. Bottom line, despite what everyone told me, these field personnel were equally impressive. They were top of the line professionals.

Well, what did I take from my schooling? First, to be a successful contractor you must employ the brightest most talented office personnel to develop, manage, and oversee all construction projects. Second, you must employ highly skilled professionals to develop, manage, and oversee projects in the field. What I realized is in order to make projects work you needed to hire some pretty hard hitters to run the office and the field. It seemed, the field put together the proposal/bid to win, and then the field compiled an entirely different method for the deliverable. Interesting, but that is kind of what seemed to work.

After just a few seconds of thinking, the accountant in came out. How does a business make money if every individual on the team is a highly compensated professional? I mean, no wonder contractors charge so much, they employ all these expensive people to perform in the office, and then they turn around and pay another set of expensive people to actually build the project in the field. There is no way that model is sustainable at least for small and medium size contractors. Also, just think of the money large contractors leave on the table by executing their operations with this structure, unbelievable!

Not to get to off trac, but in business the secret is to create a formula that does not require every step of your operation to be performed by highly paid professionals. Just take a second and think about the wealthiest companies you know, and you will quickly realize they have a lot of people working for them at a reasonable wage, not top of the line. Now before you go nuts on me, I am not saying that a contractor can run a "puppeteer business", meaning he/she hires people with a pulse and then tells them exactly what to do. No, there is a balance between hiring the most expensive people and finding those accepting a reasonable wage for their service.

Quite frankly, there is no sustainable business model that includes hiring the most expensive people you can find. The concept will not work long term, especially in construction. The reality is you hire good people and establish processes to oversee operations. Remember, people come and go, they change, they leave jobs, they get sick, processes remain in place no matter the person.

Thus far, I learned long term success is not going to come from hiring highly paid people, nor is it a matter of merely adhering to project documents, such as plans and specifications. Another system I observed is what I call the puppeteer program, which happens in a lot of small businesses. The puppeteer is the contractor that hires nothing but minions to work for him, none of them can answer a question without the contractor present, and they do absolutely nothing without his direction. In my current work, I see this type of structure quite frequently.

As with most challenging business issues, I realized the answer would not come from inside the industry. I needed to think outside the box on this, which reminded me of one of my core values, "Never act on the premise it has always been done this way." With that in mind, I started thinking about what I learned thus far and came up with two concepts:

1. Why would anyone want to run a business and leave the results up to people that might leave you for a better opportunity at any time?
2. Why would anyone trust the success or failure of a project to what one person maintains in their mind?

I began to think, what if a construction company managed every aspect of its operation through processes instead of people, would that work considering the two essential elements of project success, time and money. If I could link these concepts together, then it might be possible to manage the whole construction business as one operation. I remember one quote I read, "Finally, if a team shares a common objective, a good portion of their compensation or reward structure, though not necessarily all of it, should be based on the achievement of that common objective" (Lencioni 2012). Compensation is important, but what got my attention is the idea of one objective, one goal!

On Sunday morning, while sitting in church, I experienced an epiphany. As I listened and watched people react to what the preacher said, I questioned why all these people assembled weekly to listen to this guy. What brought this room full of different people to the church each week, and even more importantly, how could he stand up there and preach some of the things he did without anyone objecting. I looked down, noticed my Bible, and realized the reason the church worked is that everyone agreed on this one absolute, the Bible. It is the final word in the church.

I looked across the auditorium and realized that every single person in that room, all three thousand of them, had only one thing in common—the Book. Then I began to think of other religions, realizing that each different religion remained intact based strictly on one text. Then I started thinking, what is the one thing that maintains continuity in my government? A book. Name any organization, company, group, religion, nation, institution that does not have some underlying document. The document is what keeps the organization focused on the overall objective of the organization and keeps the various groups of people together. Without it, there is no true organization or establishment.

The answer to continuity lies in determining a document applicable to all parties that establishes the overall objectives, creates continuity, communicates, and addresses the two success factors of any project: time and money.

As we left church, I got a phone call from a client that ended my spirtual awakening for the moment. No niceties, just an outburst of emotion, frustration, and contempt on the other end of that phone line as I waited for a sentence without expletives. "We entered into a formal mentor-protégé relationship with a large contractor. We bid a project together and received the award, but it is not going well, and we think we are going to lose everything over this one project." The call ended with more expletives and then a plea to get on a plane and go to the project and get this resolved. I obliged.

The next morning I hopped on a plane, landed, got in a car, and drove straight to the job site. I sat down in the conference room, and shortly after that, a nicely dressed man walked into the room and said, "all of the necessary resources, planning, and subcontractors were in place to guarantee a successful project." I asked only one question, "where is the documentation outlining all of these things you just mentioned?" As every contractor across the world says, "I will send it to you by the end of the day."

Later that day, I received a large file containing drawings, specifications, the proposal, a list of proposed subcontractors, labor agreements, an estimate, and a budget. If memory serves, there were over one-hundred pages of documents to review. He did not provide a plan, but rather all of the information needed to create a plan. I spoke with my client, and she was even more agitated, so we decided to set up another meeting and see if we could get some answers.

We met again, but this time he brought all of the project team, including those people assigned to the field. The meeting became rather monotonous because every time I asked this question, "Can you provide me with a document that highlights the strategic plan for this project, including how you intend on addressing the various specifications and contract requirements?" They would, in turn, answer, "Sir, my foreman, superintendent, and QC manager respectively have over 100 years of experience...they understand the project, and when it is done, and we hand a check over to your client, everyone is going to be happy."

After a few hours of the back and forth, I gave up and advised my client not to move forward until they provided a detailed plan explaining how they intended to complete the project and manage both time and money. She thanked me, agreed wholeheartedly, and took me back to the airport.

Four months later, my phone vibrated, and I looked at the number and thought, this will not be a pleasant call. Sure enough, the voice on the other end of the call seemed a bit perturbed and ready to unload. She started, "Okay, so we did not listen to you...the project is underway. It is disastrous. The contracting folks are sending discrepancy letters, and as far as the money goes, we have no idea where we are at or where we are going to end up. I will pay you whatever you want if you can go to the site, perform some analysis, and help us get out of this alive." If I had said no, I think she might have reached through the phone and ripped my head off!

I landed and went directly to meet with the government contracting officer. He gladly explained the severity of the situation and seemed happy to give me every detail available. He expressed problems with safety, quality, budget, and the expected delivery date. (As a side note, when a project is not managed correctly, the symptoms usually identify themselves in the form of safety, quality, and budget issues. Don't chase symptoms; identify the root cause.) Before I left, he handed me a termination letter he'd drafted and said, "This is what I am about to send out. Please answer to the concerns in the letter, and maybe we can salvage this project." I requested he give me a week and then I would meet with him.

It only took about ten minutes being on-site to relate to the frustrations of the contracting officer and understand his desparation. I met with all the field personnel, and ironically each of them had a different idea of the direction of the project and spent a majority of our time together pointing out the failures of the other team members. Again, it is important when evaluating projects not to get caught up in chasing symptoms but rather work to identify the root cause.

After another hour of listening to complaints, I asked them to show me the project schedule. They took me in another room and showed me a schedule plastered up on the wall with very tiny writing. One of them said, "This is the schedule the government uses; we actually create our own two-week look-ahead deal in Excel." I asked if they had shared this with the contracting officer, and they said, "No, he only sees this one." "Let me get this straight," I replied. "There are two different schedules for this project." He responded, "Yes, there are actually three. One is for the government, one is for us, and the other we give to the suits."

I guess my questions created a stir with the large contractor because the next day the executive team from the large contractor showed up on-site unexpected. I found out later the employees of the large contractor called the office after our conversation and requested backup! After a few minutes of general conversation, the president of the large contractor said, "Wasting valuable project dollars for you to come out here and perform an analysis is ridiculous. You think coming out here and doing math is going to make this project better? We don't need all of this mumbo jumbo. We understand construction and will deliver this project just as planned." At this point, I am thinking, *"I think I know the problem with this company."*

I began by letting him know about my conversation with the contracting officer and my discussion with the on-site personnel. I further explained my concern with the multiple schedules and the lack of a consistent direction for those managing the day-to-day operations. He interrrupted me after two sentences and said, "Look, schedules are a stupid waste of time. We understand time and money and don't need analysis to figure this out. Just you being here is wasting our productive time. If you just step out of the way, we will deliver this project. I have been in construction for twenty years. I did not waste my life studying books and theories. I am out here making it happen." I remember sitting there in that chair with everyone looking at me waiting for a rebuttal. I replied simply by saying, "I really appreciate your perspective; however, I don't work for you but rather for the fifty-one percent shareholder of this project. For that reason, I am going to perform the analysis, and despite your objections we are going to make changes on this project. If you do not want to participate, then by all means, please remove yourself from our project." The room became pretty quiet. He stood up, shook his head, and walked outside to light up a cigarette. Just to let you know, not very long after, this the large contractor relieved him of his duties because they had lost serious money under his direction on several projects.

My fun began the next day as I compiled all manner of data to first determine an actual completion date. The analsysis indicated a mere forty-two percent chance for completion on the schedule contract date, and a ninety-two percent probability for a completion date pushed six months past the contract completion date. In simple terms, mathmatically this project would end up six months behind the contract completion date. I presented this data to the contracting officer and begged for a contract extension. He told me that he understood my position and would work with me on accomplishing this extension closer to the end. I certainly wish I could reveal his name because he is one of the kindest and most reasonable people to ever work for the government. Ultimately, we completed the project almost to the date I predicted. The company I represented did lose financially, but fortunately the government explained in detail on its reference form the issues with the larger contractor, giving my client credit for salvaging the job.

After this debacle, my question finally received an answer. I finally understood the "Bible" for projects. It is the only document that clearly illustrates both time and money and monitors these crucial elements from beginning to end, consistently delivering essential infomration no matter the size of the project, dollar figure, or the scope. That document is the *project schedule.*

If you currently work as a project manager or estimator you might be thinking, "Right on, there must be a project schedule on the project, I agree." But what if I took it a step further and suggested this schedule begin in the bid process and is the primary document utilized in estimating and planning for the project? Most likely, many begin to think, "Wait a minute, schedules are for management, not estimating, bidding, and proposals." But what is the ultimate objective in the bidding process? In my opinion, it is to evaluate the scope and create a detailed plan calculating both time and money for the deliverables. If managing time and money is the goal, then what better way to do that than by starting with a project schedule? Remember, the goal is not to provide an estimate but rather an actual dollar figure to complete a given set of tasks.

Bottom line, the absolute best tool for strategizing, preparing a proposal/bid, and execution is the master schedule. If you are an estimator, please hear out my entire argument before you call this heresy and burn the book. The next paragraphs explain its usefullness in managing the bid/proposal process.

Strategy

Considering time and money are crucial to the success of a project, they are integral in strategy formulation. In most companies, these two essential elements that come together to form the competitive strategy for the bid do not even come together until after award. Think about it, the estimator diligently completes the take-offs, and reviews the subcontractor bids, while the construction manager puts together a skeleton schedule. In some magical moments, the two documents reconcile, and the proposal is ready to go. As you know, that is not what happens. On bid day, the only thing people talk about is money, not time.

Before I proceed, let me clarify some concerns I hear when talking about the schedule and the bid. First, there is nothing wrong with utilizing bid software, or Excel to perform bid calculations; however, these calculations should reconcile with the mathematical analysis in the scheduling software. For example, if an estimator does a take-off and has one man working for an eight-hour day at ten dollars per hour, then the schedule should reflect the exact same thing. If the material is necessary to complete the line item, then it should be there as well. When I look at a line item in the schedule, it should identify all costs related to the activity.

One question I always get in regard to scheduling line items is, how do you account for activities that take place at different times. For example, a painter may start with a prime layer, come back and do a second layer, and then finish up with the final coat. The simple answer, add a second and third line item on the schedule. Now you can allocate the resources to each day, half-day, the painter works. It seems simple, but this is the single argument I get the most from people disagreeing with using scheduling in the bid process.

Earlier, we talked about the work breakdown structure and how it assists in ensuring everyone on the bid team understands the flow of the project. Not only did you prepare the WBS to make sure your team understood the project, but now it is the WBS to start building the project schedule. With the right software, you can always filter the line items into the WBS structure. Breaking down the features of work helps everyone to understand how all the features of work come together to complete the scope of work. If the entire team works off the schedule, they can refer to it throughout the process, observing how it comes together, the utilization of resources, the identification of risk, and ultimately the crucial relationship between time and money.

For illustration purposes, think of the strategy aspect of scheduling as similar to taking a trip. Say I am a client, and the deliverable I request is a trip to Pensacola, Florida, from Las Vegas, Nevada. The destination is the scope of work, which includes the plans and specifications. The first step in the process is to identify the work breakdown structure, or the intended path of travel. The entire team works on this, ensuring that the most efficient travel route is defined. Second, the team begins adding detail to the schedule, evaluating line items individually and determining the cost of each specific activity. This information is readily accessible to everyone on the team, and it keeps the crew cognizant of the cost. At any point during the bidding, every team member knows the time and money involved with the project. Also, with it broken down, team members identify risk per line item and address those concerns by adjusting the pricing to compensate for risk. On bid day, the schedule provides every detail, including the strategic plan, resources, and cost to deliver the scope, or in this case get me to Pensacola, FL. The schedule is the one document bringing both time and money together.

Now, using the same illustration let's play my trip without a schedule. First, the scope is evaluated, and then the travel team begins working on the best route while the bid team works on pricing. The travel team figures out a great route to get me to Pensacola in the most efficient manner. Meanwhile, the bid team works diligently on figuring out the cost of the trip. Think about it, the bid team is figuring out pricing without really knowing the direction selected by the travel team. Not a lot of strategy going into my trip because the money team and the time team are working separate from one another. Do you see the apparent difference? If nothing else, at least recognize the importance of the collaboration and consider what happens if awarded. The first group hands over a fully defined time and money plan to the field team, and they deliver a promise to a customer. The second team hands over the plan to the field, but the time and money don't line up; therefore, the field team creates an entirely new strategy.

Schedules also enable the team to reconcile time and money when it comes to resources. Without a schedule, the majority of evaluating resources such as subcontractors, vendors, and suppliers is done solely based on money and comparisons to other bids. What if your team looked at the schedule line item and realized the labor price did not line up with the time in the schedule, wouldn't that be great to identify now instead of later? Evaluating resources through the reconciliation of cost and time enables the bid team to identify economies of scale or to combine resources to reduce project overhead.

I remember one project in particular that involved a renovation of a gymnasium. The team working for me at the time became pretty good at this process, maintaining about an 8% kill rate. They were very successful because they evaluated resources through the utilization of the schedule and the budget. On this project, they identified a common denominator with the various resources; all of them were bringing man lifts to the site. When they looked at the schedule, there were about fifteen days where lifts were delivered to the site or picked up from the site. The total cost for all the lifts ended up at about $25 thousand. By identifying this through the understanding of time, they decided to rent two lifts for the entire project for a total of $10 thousand. The time along with the money helps in making the right decisions.

Risk assessment becomes critical as strategies become a reality. When a project schedule is utilized in the bid process, each line item stands alone and enables personnel to consider it by itself and with other activities from a risk standpoint. This also becomes important during the final stages of the bid. Instead of just tacking on a random percentage of every single line item, a team can assess each line item and determine its individual risk. This simple process might add that cost advantage necessary to win.

Another great thing about scheduling is the separation of the project into phases, line items, or activities. Construction is difficult to understand sometimes because of the numerous activities going on at one time. With a schedule, you can break any part of the work down into a few line items, which helps explain the process. More importantly, it assists in identifying risk or developing strategy. You want to understand how to complete a task efficiently break it down into little pieces.

Think about it this way. If I said to you, I am from a different planet and getting ready in the morning is not something I am familiar with, would you provide me with what I need to do. You might say, sure, get out of bed, brush your teeth, comb your hair, and get dressed. If I followed those steps per your instructions, do you think I could get ready? No, of course not. I would need to know what material I needed, what about how to brush my teeth? If you want your team in the field to execute, you must provide them with the details on how to accomplish the task. If there is a great strategy, I want to make sure the team carries it out to the details.

Of all the words in the English language, I believe implementation is the most difficult.

Humans suck at follow-through. If you are a contractor, you know this is true. Think about it, what is the most challenging part of any job, it is the last few months with closeouts, submittals, and all the crap you should have done months ago. A lot of people are great at making plans, but successful people carry those plans out. In 2008, I completed some consulting work for a small contractor working on a Navy project. He became very frustrated because he got tired of seeing the actual cost vary so much from the bid. When he first brought me in and asked my initial thoughts, I said, "Most likely, the great planning you do in the office never makes it to the field." He agreed and then decided to pay me to go out to the field and get some answers. Despite the fact, I gave him the solution for free; I let him pay me to visit his job.

I visited four different jobs and interviewed all of the team on each project. While I interviewed them, I had on my computer the plans (bid schedule, strategy, subcontractors, schedule) the bid team put together for each project. I intentionally asked the field personnel questions to see if they knew anything about the bid strategy or the plans for the project created in the office. Without exception, all of the team members had never seen any of the documentation I possessed. Most of them arrived on-site, developed a new schedule, hired new subcontractors, and did what they thought worked best for the job. I went back to the office gave him the same answer and let him pay me this time.

One final note on schedules as they relate to strategic planning. When done properly, schedules produce logic, which leads to the creation of a critical path, PERT (program evaluation and review technique) analysis, slack information, variance information, and expected completion times. Take a minute and pull out the old quantative analysis textbook from college and look at these calculations—pretty intense information, However, if solid scheduling software is utilized, all of this information is available at your fingertips during the evaluation of strategic approaches. I am not sure about you, but if I am putting my own money on the line I want every tool available to ensure my project goes off without a hiccup!

Preparing the Proposal/Bid

Construction is an odd industry when it comes to personnel. When you work for a construction company, the level of expertise among staff varies quite significantly. For example, the estimator might be a former field person with vast experience and knowledge of the trade work, but little understanding of construction controls. I find this to be true, especially when it comes to scheduling. When you talk about a construction schedule, it means something different to people. To some, it is a quantitative tool or a network analysis, and to others, it is nothing more than a planner.

Terms such as critical path, slack (float), and sensitivity analysis are not just words used for scheduling; they represent mathematical calculations. I remember a particular case I worked dealing with a delay analysis. I met the contractor and discussed the delay claim with him in detail. He threw out all the right terms and seemed like he understood scheduling until I started asking questions. After a few minutes of conversation, he told me that the critical path is the plan the contractor implements to finish a project on time. At first, I thought he meant something different, but he did not understand the critical path is a math calculation and is not controlled entirely by the contractor.

As I said, working with professionals, I assumed they understood the math behind a critical path method schedule. On one project, I dealt with an individual that by all measures, is a bright guy. His frustration with scheduling was that the critical path developed by the logic-based software did not match the path he thought would work most efficiently. At first, I thought his frustration resulted from a lack of entering data correctly, but after further discussion, I realized he wanted to create a critical path. Of course, this meant he did not understand the mathematics behind the critical path. My point is these types of calculations are mathematical, providing another level of expertise in the never-ending effort to identify actual cost during the bid process. Remember, we do not seek an estimate but the real cost. After award, we do not have the opportunity to tell the client our "estimate" came in low. That is not the way it works.

Here is the point, schedules provide a valuable tool for contracting. First, they assist with understanding the two most important concepts for completing a project, time, and money. Second, when it comes to change orders and delay disputes, the schedule becomes invaluable. Third, just like budgeting provides a financial plan, scheduling provides a project plan.

Consensus accomplishes nothing. My number one pet peeve is when people agree to avoid discussion. You know what I am talking about because we all do it. Enter a meeting, here an idea, think it is the most ignorant idea ever, but shake your head and tell the presenter it is excellent! When it comes to the project schedule, it should be a full-on argument with the team. Everyone thinks, goes through the details, and questions the activities, the sequencing, the resources, the logic, and every aspect.

Scheduling is a controversial topic, believe it or not. Unfortunately, big and small contractors ignore this valuable tool. I cannot tell you how many jobs I visit where the schedule is prepared by a third party to meet a specification. I worked with an individual, a guy I hold in the highest regard, and we argued for years over the value of a schedule, particularly in the bidding process. Despite his objections, he finally incorporated it into the bid process and later admitted that the scheduling factor had a significant impact on the bid. After you read all of this, give scheduling a try, make it part of your process, and I assure you it will be worth it.

Putting together a price to complete a task with millions of unknows is challenging. Estimators incorporate tools such as reviewing cost on past projects or calculating material and labor from piece rate calculations or depending solely on input from a subcontractor/vendor. As an accountant, I understand the approach. After all, in the investment world, decisions about the future are made from past indicators. No one is arguing the value of historical data; however, just as with finance, the past is no guarantee of the future. Unfortunately, the current system for developing bids frequently only evaluates past information or cost without consideration for current market conditions or costs. The thought patterns permeate the process. For example, the estimator asks for a bid, the subcontractor supplies this bid, which falls within the historical range assumption made by the estimator.

It is classic groupthink where because several people agree that something sounds or appears logical, it must be correct.

Okay, follow me carefully. Estimators typically work in a vacuum. They compile a lot of information, input it in a spreadsheet, make assumptions, and then generate a number for bid. It seems logical, right? No! Why would I as an owner, knowing there are millions of variables or unknowns, trust just one person to provide a number. How do you know it is accurate? What if he missed something or made a wrong assumption? The truth, most contractors, trust the individual and in blind faith hope for the best outcome. Therefore, you trust your entire bid to this one individual and his/her complicated spreadsheet. Unfortunately, we depend on estimators way too much because we have no way to quickly evaluate their hypothesis on the cost of completing an activity.

Before you start thinking I am on a mission to defame estimators, hear me out. Please take a minute and think about the structure of our government, in particular the military. When it becomes necessary to perform a strike, take aggressive action, or go to war, who is in the best position to make that decision. We all agree that the military is the best to execute the mission, but letting the military decide to proceed means every conflict is solved by force. What happens then, the military prepares a detailed plan and gives it to the executive branch. The executive branch reviews the plan and makes the final decision. In estimating, the estimator may prepare the numbers, but the rest of the team needs to make the decisions.

How do you do that? Glad you asked. The answer, the project schedule. When there is a schedule available, it breaks down the project into activities. The team can review the activity, the resources, the time, the material, and even how it sequences with other events. Each team member can begin to ask a question specific to an activity, or about a resource. The estimator just got a whole team to review and understand the entire bid.

My company pursued an Air Force project requiring both in-house resources and the use of subcontractors. The team followed the process as presented in the book and prepared a full schedule. The job, not very large, required only one estimator. He put together the bid and incorporated it into the master schedule for review by the team. On the day of the master schedule review, we had one of our laborers in the office doing some renovations. I asked him to join us while we reviewed the master schedule because I needed the noise he made to stop! The scheduler posted the project on the big screen, and the team began discussing the line items. The kid I asked to join us blurted out, "How come the labor is so high on line item A2320? If only takes two days to complete, why is it so much money, I need to do that work." The estimator completely ignored the kid and kept on with his explanations. As I started to read the line, the kid shouted again, "how in the world can you charge that much money for two days worth of work?" By now, everyone started reviewing the line item. The kid caught the error even before we did. The estimator incorrectly calculated the labor increasing the line item price by approximately forty-two hundred dollars. There were thirty-two line items with the same work, which equates to one hundred, thirty-four thousand dollars. We adjusted the numbers and ended up winning the project by about twenty-five thousand. Without the schedule, we would have overbid the job and never even knew why.

One more quick story...My team pursued a project in California that depended solely on subcontractors. My team prepared the WBS, entered all the activities required to complete the work, and reviewed subcontractor bids. We started comparing the bids to the time in the schedule and quickly identified several subcontractors charging more labor than the time involved to complete the project. We called the various subcontractors, explained what we found, and believe it not, each of them made adjustments to their pricing. Again, we did win this job, but if we had not performed this level of review, it would not have happened.

Construction is full of assumptions and contingencies, making it a subjective pricing game for bidding. There is a similar industry that bids on the future, insurance companies. The difference between insurance and construction bidding is the insurance companies embracing tools and data in the bid process. Insurance companies assume risk just like construction companies. Consider a life insurance policy for half a million dollars. Each month the insured pays out eighty-dollars for this policy. If at any time in the next twenty years he/she dies, the family gets half a million dollars. Think about it, if you die within two years, you will have only paid one thousand, nine hundred and twenty dollars, but your family gets half a million. Risky for sure!

The insurance company hires actuaries, develops detailed statistical models, studies trends, and mathematically predicts the details necessary to ensure profitability despite the risk. Insurance companies utilize every resource available to them to ensure profitable operations. Construction companies try and get away with as little tools as possible. They do the calculations in their head, hope for the best, and keep putting numbers out there. If you want to make money, then do it right, use the tools available to put together accurate bids.

Let's walk through an estimate incorporating scheduling. Instead of thinking of the estimating process, consider what your mind does naturally. You begin reading the scope of work; your mind immediately organizes the work into activities. Go ahead, try it. Your mind will automatically start processing what you must do to accomplish whatever project you imagine. Again, with no effort, the brain then begins to add time to the activities, how long it will take for you to complete the objective. If you still do not believe me think about what your brain does when your wife gives you a list of things to do, and there is something else you would rather be doing.

The human mind operates like a computer, and as much as you may hate math, your brain uses it all the time. As you start reading through the scope of work, the brain does all of the things I talk about naturally. The issue with the mind, or at least my brain, is that it also does a lot of other things and quickly removes information that it deems unnecessary. Contractors get in trouble when they operate under the guise of keeping information stored in their brains.

Earlier in the process, I mentioned the WBS (work breakdown structure), or the outline for the project. The WBS is so essential for understanding the project and preparing a strategy for completion. Let me illustrate the purpose of the WBS by relating it to something a bit more practical. My daughter, the smartest and most beautiful person in the world, is a junior in college. From time to time, her studies require a research paper or literature review, which means she needs to prepare a ten to twelve-page paper. The first few research papers she wrote, exemplify what new students do and what professionals do when it comes to writing. The research paper is due on Friday, on Monday, with the whole week before her, she gets her computer out, sets up the formatting, and then starts typing. I refer to this method as divine writing. She will put her fingers on the keyboard, and God will take care of the rest.

If you ever wrote anything, you know that is not how it happens. To write anything, you need to do a lot of reading or preparation, and then you prepare an outline to organize your thoughts. Once the outline is complete, then you can start writing. The research for projects comes from the specifications, the contract, scope of work, and other relevant material. The WBS is the outline of the research and establishes the foundation for the job, the project.

Let's walk through using the schedule in the estimating/bid process. At this point, the team has prepared and agreed on the WBS, which establishes the project's overall outline. Next, the scheduler, in coordination with the other team members, adds the line items or activities required for completion of a task or deliverable. The more details, the better. Remember, this is the field's information to carry out the overall strategic plan for the project. Guess what the number one fundamental of strategy execution is, making sure people know what they are supposed to do. The schedule gives field personnel direction, which ensures the strategy created in the office is executed in the field. Same deal with subcontractors, if there is a certain way you intend on things being done, make sure they understand the plan. For the record, include critical subcontractors in the scheduling process, the buy-in is always helpful.

As the activities come alive, the team begins to see the project coming together. The plan starts to take shape, and the team identifies risk, holes, or even areas for improvement as a result of overlapping or structure of completing work. When the projects, activities, come to life in a schedule, people start understanding expectations. For example, they understand the flow of submittals, the timing of approvals, the impact of delays, and the importance of communicating with the entire team. Even non-construction personnel understand the flow and many times bring up some great points for consideration. All of that with just one tool.

With the activities added, the team begins entering the time required to complete each activity. This is a crucial step and requires the involvement of every team member because this is one of our key factors (time and money). This information sometimes comes from subcontractors, or internal experience, or some other analysis. This is much simpler because it is broken down into activities, so instead of attempting to assign guesses to how long an entire feature of work takes, one can just add all the activities and know the time. Before you get too excited, I realize that many tasks overlap and have different relationships, which we will discuss shortly. Another significant element of doing it this way is everyone participates in determining the time to complete a task. This is where a lot of debating takes place, ultimately leading to accurate data in regards to time. Remember, successful project management means managing both time and money.

After entering the activities, the team begins reviewing each line item to allocate the necessary time. Most of the time, the scheduler enters the time along with the activity, but I am separating it for explanation purposes. Allocating time to activities is just as crucial as the logic of the schedule (the relationships between activities). The time allotted to an activity line item represents the cost driver. For example, if the activity is for one day or eight hours and requires one laborer, the cost is the laborer's hourly rate times the cost driver, hours. The important thing here is the team can look at the cost driver and make sure it aligns with the actual cost. The entire team evaluates activities instead of just looking at the big picture and saying, "the number looks good to me." Hopefully, in this phase, there is lots of debate about the cost and time. The more the team argues, the closer the team gets to the right amount or time on the line item. Again, the two most essential elements of a successful project are time and money.

Great news! After you run through the schedule, adjusting the time, you just cracked half of the equation - time. The more eyes involved, the better. It always amazes me what people see when they get involved and study the schedule. As planned, you end up with a win! One thing I do want to point out is the scalability of the schedule review. If the project is enormous with a million line items, then probably separating it by WBS and reviewing with separate teams works better.

Now, the real fun begins. The schedule has logic (simply put - the scheduling system ran network calculations), and resources or responsibility codes indicate who is doing what. Right about now is when my estimator friends start getting upset and pointing out all types of flaws with using a schedule in the bid process. In most scheduling software, there is a separate resource page allowing for set up of the resource, the cost, and the driver. The resource might be a person, or a title, like a project manager or superintendent, or if it is a Davis-Bacon job, it might refer to a specific trade. The point is when you add the resource to the line item, the system does the calculation. Estimators do not like this because it does not match the number from the separate estimate, which is precisely the point of the exercise. Why wouldn't they match? Eight hours times, the burdened rate should equal the same no matter what software applies. If you intend to use a subcontractor, just put their total in the line item, or you can break it down. Straightforward system and allows an instant checkpoint.

Every activity on the schedule must include a resource. Most include multiple resources explaining what is required to complete that line item, who is responsible, and the cost. All of this is in one place for the entire team to evaluate, discuss, revise, and put into the final draft format for the bid/proposal. No complicated spreadsheets are necessary. The process does not lean on just one fallible human being, but rather a document giving a clear view of the two most essential components of a successful project—time and money.

When all the project data is in the schedule, there are literally a thousand different sorting options to assist with understanding the bid. How about sorting it to show labor? Or what about by resources? Or how about in a format that illustrates the flow of money or the budget for the project? All of this information is available at your fingertips before the bid is released, enabling the team to complete just about any analysis imaginable. As an owner, you can now show up on bid day and understand the project, the strategy, the flow of work, and the final number. With a few clicks of a button, you can ask questions that might be the difference between victory and a second-place finish. The utilization of the schedule in the bid process is priceless!

When the schedule is complete, resources added, and costs calculated, the last step is to review the total cost to deliver. Remember, at this juncture, the total cost does not include profit. Everything else is on the line items, except for profit. Think about it this way, the number on the schedule represents the cost; after you add profit, you have the price or the total contract value.

What I am going to explain now is probably the number one reason contractors lose work. Unfortunately, a majority of contractors bid on a cost plus a percentage. If the cost is one-hundred dollars, they multiple that time ten-percent, and the contract value is one-hundred and ten dollars. The problem with the method is everyone does it the same way, meaning the company that bids differently gains a competitive advantage.

My company maintained a high kill rate (projects won/projects pursued) based on what I am about to show you. You want to win projects, implement this step immediately, and your win percentage increases. Every line item (activity) on a schedule represents a risk. Profit from a project comes from the risk/return theory, meaning the more the risk, the higher the return. The line items in the schedule represent different levels of risk; therefore, when determining the return (profit), the contractor should evaluate the line item and consider the associated risk. Instead of just tacking on a random percentage, evaluate the line items, and adjust the profit based on the specific activity's risk.

How about a real-life illustration. The government does some funny things when it comes to contracting. My first company, Power Services, worked at a base in Northern California that needed fifteen Caterpillar generators, the big boys. For whatever reason, the contracting officer could only do this by issuing fifteen separate contracts for one hundred thousand dollars each. There were eight contractors eligible to bid, and we all had the exact specifications, labor, etc. The specifications included preparing a site pad, removing some trash from the location, and contracting with an electrician to install the connection point. Caterpillar did the rest.

Power Services won all fifteen of the contracts. The other contractors were furious, but after they calmed down, we talked, and I knew why we won. They compiled the cost and marked the entire package up by fifteen percent. In our bid, we analyzed every line item and calculated the markup based on the risk related to the line item. For example, I had zero risks associated with the generators, none at all. Caterpillar delivered them, set them on the pad, and performed the connection. I had terms for payment of sixty days, which is plenty of time to submit my invoice and get paid from the government. No risk! I bid the job, reduced the return percentage on those line items, and won the job with a twelve percent margin. Markup line items based on risk, it will get you in the winner's circle!

Execution

Earlier I mentioned the disconnect between the field and the office. Project strategies, created in the office, never make it field construction office. In some cases, the superintendent and project manager get called and told to go and do the job. They take the information and start the whole process over again, and then execute an entirely different plan. Both parties follow the specifications, but sometimes the cost associated with making a new plan erodes the original margins. When the bid team follows the procedures as outlined, they provide to the field what I refer to as a "project in a box."

In the early part of the book, I addressed one of my observations about large contractors and how they successfully managed projects. I noticed they hired some pretty heavy hitters to not only prepare the bid but also to manage field operations. They resolved disconnect issues by placing top-line professionals in every aspect of the business. Of course, because the jobs are so large, they can afford to do that and remain solvent. If you are a small or mid-size contractor, you probably only employ one or two top line people because that is all you can afford. If you did hire the high-priced personnel, it would shrink your margins and probably squeeze you out of business.

The project in a box theory is what I do to overcome the issue with high paid personnel. In our company, we invested the time in the bid process, which included building a detailed schedule with step-by-step instructions on how to manage and complete the project. My team did a great job of lining out step by step instructions to the field. The schedule, along with the other contract documents, provided the field personnel with everything needed to deliver the project according to the plan. The project manager made sure the field personnel followed the outline of the schedule. Even the money side of things got handled by the schedule. For example, when the superintendent needed to order material, he requested a purchase order from accounting using the line item from the schedule. The accounting staff would issue the purchase order to the superintendent as long as the request did not exceed the amount on the activity or the line item. If it exceeded the amount on the line item, the project manager got involved and determined the issue or challenge. The one document managed the entire job. Nothing is left to chance; you must manage both time and money.

People are always an essential part of the equation, but if you overpay or only employ highly compensated individuals, that limits your competitive capabilities, especially on small to mid-size jobs. If someone does a great job on a project, give them a piece of the pie, incentivize them to follow the plan. If the bid team develops a detailed schedule, including step-by-step instructions on how to carry out the process, then instead of paying top dollar for field personnel, there is an opportunity to hire less experienced individuals capable of following directions and completing the project. The project manager that worked on the bid team oversees the project and works directly with the less-expensive field manager, adding money to the project's bottom line.

I love watching college sports, especially football. Unlike professional sports, many of the athletes play strictly for the love of the game, not for money. When I lived in Florida's panhandle, I attended as many games as possible. I met the owner of a local radio station that did interviews with the coaches and players at the University through some business dealings. He had worked in this role for years and became good friends with the coaches, which resulted in invites to a lot of cool events.

Not sure how I got the invite, but my friend asked if I would join him for a home game. Not only would I get to watch the game, but I would get to go behind the scenes and see how it all comes together for the big day. Upon arrival, we toured the facilities, and I noted that most of the activities related to the game dealt with planning. There were tons of rooms where players and coaches did nothing but talk about the game, strategize on how to win, discuss risk, and ultimately put together in writing a full game plan. Once they agreed upon a plan, everything that happened the rest of the week, including the practices and drills, revolved around executing the plan.

The game obviously represents the project you intend on pursuing. The coaches include the team developed to put together the winning strategy, and the players represent the field personnel, subcontractors, vendors, and other stakeholders. The coaches invest hours of time in going through everything related to the game to make sure they understand both the opportunities and the risk. They meet multiple times and finally draft a game plan, a schedule, explaining exactly what every team member must do to win the game. There is nothing left to chance, they run all types of scenarios to ensure the plan works. On game day, the coach, project manager, along with the team execute the plan. Do they make changes, of course, but ultimately the detailed plan is what enables them to make the last minutes alterations or change a play on the fly. Millions of dollars go into planning for a football game. Doesn't it make sense as a contractor putting your livelihood on the line, it would be important to learn something from football. One more thing, the players on the field are just a bunch of kids, not professionals, just eighteen year old kids fresh out of high school.

In my line of work, I meat all types of contractors with a range of skill sets. Great business people, technical experts, influential leaders, and those I am not sure what they do. No matter the individual strength, the successful operations all have one clear denominator, a clear and consistent direction with an understanding of personal limitations. Let me explain how this relates to the process.

Construction is full of distractions. Spend one day with a contractor and observe the phone calls, the issues, and employment issues. It never stops for a contractor. To keep a clear and consistent direction with all the noise is a challenge. Many contractors try and manage everything in their heads, meaning every question, issue, problem, or opportunity requires the contractor's involvement. When you follow the procedures outlined, the people that work for you do not need to harass you because there is a plan in place they can reference.

To illustrate, let me give you the rundown on bidding a government project. The government releases all of the documents and establishes a site visit. The contractors show up at the visit, and the representative tells everyone to submit any request for information in writing within forty-eight hours. All of the contractors then prepare a list of RFI's and submit them to the government representative. Ninety percent of the time, the government response to the questions is "answer included in provided government documents." Guess what, all those questions you spend hours a day fielding, are included in the documents!

One question all of us hate is, "What do you feel is your greatest weakness?" What a question. Most people answer with some lame response just to make themselves look good, "I trust people too much." One struggle we all experience is none of us know everything. No contractor knows everything there is to know about construction, nor does anyone have all the answers.

As you read this, you might be thinking about your weakness. Also, you might compare this weakness to struggles within your business because, as you know, the "speed of the leader is the speed of the gang." If you maintain technical prowess but lack in the contracting side of the business, a schedule can help you understand how to address items that don't particularly attract your attention. Also, you can assign these task to other team members or a third party to ensure it is done correctly, making you look like a great contractor. After all, it is sad when I see a superb technical contractor perform poorly based solely on his lack of paperwork skills. Utilize the resources available to overcome what you do not know!

On the other side of the coin, there is a growing field of contractors that understand compliance and paperwork but lack technical expertise. If that is you, the schedule will give you insight into construction like never before. I could tell you story after story of using a schedule to argue with a technical expert by just applying a little common sense. Again, utilize the resources available!

Oh, did I mention that a schedule with each line item costed out also provides a perfect schedule of values to guage the project completion and the payment by the client. The schedule really does function as the "bible" on the project. It is a great tool for managing every aspect of the project from cradle to grave!

The idea of using a schedule to manage the bid to build process is something I firmly believe is a necessity in an industry of unknowns. The schedule provides so many tools to both large and small contractors keeping them solvent, out of court, and understanding where they are at on a project, and how delays impact the money and the deliverable. If you still are not convinced, I would merely ask you to give it a try. Next time, start with the schedule in the bidding process and let it work for you!

How to build a schedule

I included a section on building schedules because I know a lot of contractors avoid tools because of the complexity or the difficulty in incorporating it into operations. Scheduling does require effort, but I assure you the benefits outweigh the cost. Besides, it does not take very long to master scheduling no matter what software you decide to utilize.

To start, choose software that includes logic, resource management, cost loading, and network analysis. I recommend Primavera. I will warn you; it is tough to set up and requires some training. The beautiful thing about Primavera is there are companies out there that will host for you, which means you access it through the web, and they keep up with all the other fun stuff. Lots of contractors use Microsoft Project, which is fine if you manage it correctly. My only issue with Microsoft Project is the ease of manipulating the schedule when a logic issue occurs. Just make sure whatever scheduling software is selected, you let the logic do its job.

Together, we are going to walk through the steps of preparing a schedule. If you are a scheduling expert, please understand I am trying to make this as practical as possible for a novice. If you keep working at understanding schedules, you will improve quickly and implement all kinds of features that assist with the oversight of projects. There are excellent resources available on preparing schedules. Please, invest some time in understanding the power of the scheduling tool.

Step 1: Develop a WBS (work breakdown structure) for the project.

Throughout the book, I discuss the importance of the WBS (work breakdown structure). To me, the WBS is my breakdown of the project, and it is how I break the project up in my mind. When you work with a team, everyone has input, but at the end of the day, the WBS gives the general outline of the work for the project.

A lot of schedulers utilize CSI divisions for WBS. In some cases, the owner may request a WBS based on how they see the project. There is nothing fundamentally wrong with sorting a project in either of these manners. The beautiful thing about scheduling software is you can sort activities in about twenty different methods. To me, the WBS, belongs to my team. It is how we see the project flowing. If the owner wants to see a different flow, I still start with my WBS and add their flow to activities and sort it accordingly.

As I illustrated earlier, if you intend to deliver a speech, you start with an outline of the points for discussion. Same deal with the project.

Step 2: Add Activities under each WBS section.

Normally before I add activities, I set up various codes (a term in Primavera), so I can track my line items differently if necessary. For example, in some proposals the owner might request you provide the schedule in accordance with the CSI divisions. In that case, I would set up a code for the various divisions and as I enter activities or line items, I would add the code to the activity. By doing this, I can sort the activities in whatever format necessary.

I did not say this earlier, but it is okay to add sub sections under the main WBS headings. Again, it is your outline. Set it up based on how you see the work flowing. Once done, start adding activities. When I start adding activities, you might think I lost my mind because I talk out loud. I add detailed activities because I prepare it like the person that is reading it has never done this in their entire life and I must break out every single detail in order for them to do it correctly. Do not worry, if you forget something just add it, you can address the order of operations when you add the logic.

One of the smartest people I ever worked with struggled with schedules because he skipped a lot of steps. I worked with him for months trying to get him to understand the importance of putting each step of whatever task in the line items. I finally decided that talking about it would not work, so I decided to give him a real-life illustration. I told him to write out a detail of how he gets ready in the morning. He put together the generic line items, brush my teeth, comb my hair, get dressed, drive to work. Once he had it all on paper, I told him the next morning to call me before he got out of bed. When he called, I said, what is your first step in the schedule. He said, brush my teeth. Before he said anything, I responded with how you plan on brushing your teeth if you have not even got out of bed. For the next half hour, I did this with everything he wrote on the schedule. He got the picture, and from that point on he added details. You can always remove or add lines with a quick push of a button.

While working on the schedule, make sure all of the contract documents are available for review. As I add line items related to an activity, I like to peruse the contract documents to ensure I include everything required. Construction requires careful planning, meaning you do not just show up one day and complete an activity. There are usually a lot of things that must happen before you ever actually complete work. For example, when we did work for the government, specific procedural steps took place before we began work. If you did not follow the process or include it in your schedule, the project would come to a halt while you waited for the government to approve the paperwork.

Even the best of schedulers forgets to add all of the line items. That is why there is a team in place to review the schedule per the WBS everyone created. When presenting the schedule to the team, each member will understand the process and point out missing line items. The team will assist with ensuring the line items provide enough detail.

Step 3: Add durations on the line items (activities).

As you become more efficient at scheduling, adding time is probably something you will do as you enter line items, especially if you know how long something takes to perform. Some people only do this after adding the line items, while others do it simultaneously.

Durations are struggle points for those preparing schedules. I have reviewed schedules that include fifteen to twenty extra days for weekly site meetings. On one particular job, the scheduler included the weekly meeting for the two-year project adding one hundred and four days to the schedule. Here is the simple fix, if a line item is not going to interfere with the actual construction, but is a necessary task, then just put zero for the duration. It will show up on the schedule but has no impact on the project's critical path.

On durations, do not forget that resource allocation impacts time. If an activity requires one man for eight hours, what would happen if you had two people for four hours? There are a lot of variables for consideration on activity durations. When entering line items or activities completed by subcontractors, find out how many days it will take them to complete the activity. Enter that time, so the schedule will reflect how long the individual scopes of work last.

If your company self-performs work, then you should have a pretty good idea of what resources (how many men) are available each day on the project. Remember, when you add resources (costs), the system is going to multiply time and resources and then add material to show you the total cost for the line item.

I cannot stress how important it is to take time to go through the durations for the various activities. Remember, time is money, so if you can reduce the time for activities, you reduce the time required to complete the contract. In competitive bidding, managing the time to complete the work is crucial to submitting a winning number.

Step 4: Establish the schedule logic.

Again, some people do logic before time, and others do time first. I do both logic and durations at the same time because it assists with understanding the flow. Logic is where the math or network analysis comes into play; it is where the software performs behind the scene calculations and provides lots of valuable information as it relates to time. Before we go any further, let me give you the options available in most scheduling software for relationships.

1. FS (Finish to Start): By far, this is the most popular selection for relationships. It simply means, that before beginning an activity, the previous activity must finish.
2. SS (Start to Start): This is one of my personal favorites, which I think is not used enough. It means that two activities can start at the same time, or if you add some lag to one activity, the second activity can begin shortly after the first one began.

The final two are not as popular as the first two. Quite frankly, if you work on using the first two to begin with, you can incorporate the second two as you become more comfortable with relationships (logic).

3. SF (Start to Finish): This relates to a handover type task. For example, in order to finish my work shift, the replacement must show up.
4. FF (Finish to Finish): These addresses activities that occur in parallel but cannot be completed without each other.

For the software to perform calculations, every line item must include both a predecessor and a successor. Take the necessary time to think through each activity to ensure the linked line items (activities) are indeed related. I have seen a lot of contractors get themselves in a tight spot with the Corps of Engineers because of non-related activities causing the schedule to appear delayed. The keyword to remember is logic. All relationships should logically connect.

As you enter the activities, constantly push/click the schedule button to ensure what you just entered makes sense. Most scheduling software allows you to make adjustments and then push/click on a button that performs the math. Do this often, the last thing you want to do is enter a massive amount of data and then find out there is a loop (relationships do not make sense) in your logic.

After entering all of the logic, there should only be two-line items without predecessors/successors, the project award line, and the project completion line. Entering relationships sounds so simple, but most of the time, when I review schedules, this is the first issue I find, line items that with no connection to other line items. Without the relationship, the system cannot do its job and perform the calculations for you.

The logic should make sense. When I prepare a schedule, I talk through it and engage other people on the team to assist. Involve the team; remember this is a joint effort, and just because you might be the one preparing the schedule, does not mean the other people cannot contribute.

Step 5: After you finish step 4, the schedule should provide an organized flow of work giving you a start and completion date. This is an excellent time to do what I call the "common sense" review. The common sense review includes everyone looking over it to make sure that it makes sense.

I cannot tell you how many times professional estimators skip this step and submit a schedule with something crazy like the completion date starting before the project begins. I included this as a step because it amazes me how many people finish doing a schedule and then just walk away and assume it is perfect. Make sure you check your work, going week by week through the schedule to make sure it functions as designed.

Step 6: Add resources to the line items (activities).

With all the activities and relationships defined, the first draft of the schedule is almost complete. The next step is to add resources, or who or what, for completing the activities. Again, as the scheduling skills improve, entering resources might occur along with the other steps.

First, every line item on the schedule must include a resource or a responsibility code, depending on the software. Think of it this way if it helps, every activity requires either a person, equipment, material, or a subcontractor for completion. If there is no resource, the activity is not getting done. In construction, resources include labor, materials, equipment, subcontractors, owners, vendors, government agencies, architect/engineer, quality control/quality assurance personnel, and suppliers. Most line items will include several different resources, all working together to complete an activity. There are typically internal and external resources to consider in preparing the schedule.

Let's begin with a discussion of internal resources. Internal resources include line items the company that prepared the schedule intends on performing or self-performance. Follow the scheduling software directions, but the objective is to add the resource and let the system calculate the dollar value. For example, if there is an activity scheduled for two days with my in-house carpenter, then once I set my carpenter up as a resource and add him/her to the activity, the system will perform the calculation and provide it the cost for the line item. If I need to add two carpenters, I select two different resources. To add material or equipment, choose it from the resources previously set up, and all of this is added directly to the line item.

At this juncture, people ask me about calculating material as it relates to self-performance. Advanced scheduling software does possess the capability to do calculations for material, but I prefer to let the schedule deal strictly with time, not linear feet or no. of units. What I do is add resources as a line item and then attach my spreadsheet or quote to the line item so the team can review it when looking at this line item. Please include the name of the resource, so the field people know the company selected to provide the material.

Once the subcontractor provides the activities to enter into the schedule, the next question is the dollar value of each activity. The detailed information acquired from bidders (subcontractors, suppliers, and vendors) is crucial in making the right strategic decisions for the project. Ask questions, understand the bid, and communicate with the external parties on how they intend to perform on the project. Ask them to break the pricing down by line item so that you can enter the cost in the schedule. None of them like doing it, but the information or numbers they provide is how you will build the schedule of values later in the process. Remember, if the subcontractor is difficult to deal with in the bid process, I can assure you they will be difficult in the execution of the work.

Sometimes, as anyone knows that works in construction, the bids come in at the very last minute. If that is the case, go ahead and have a TBD (to be determined) as a resource. When the team reviews the schedule, everyone will quickly identify the deficiencies and work collectively on a resolution. For those last-minute bids, you plug in the number you intend to use in the schedule and can address the breakdown after award.

As the information becomes more available, you may need to adjust the times originally entered. If so, all you do is increase or decrease the time and then have the system schedule. It immediately adjusts the activity and the entire schedule to reflect the changes. A good scheduling system maintains all this information for each line item. This means the resources on the schedule are the exact ones used in the bid process, which really assist with those taking over the project after award. Just think how valuable it is for the field personnel to understand the strategy of the office team that prepared the bid. Field personnel then can attempt to control resources because they understand the master plan.

Step 7: Sorting

The last step in schedule preparation deals with the presentation. Projects include multiple stakeholders with varying interests. For example, subcontractors only care to see the part of the schedule impacting their work. The owner is interested in sorting for payment and progress, and the quality control team wants to see the breakdown of work to manage inspections. On the internal side of things, the team might want to review the self-performance or look at the critical path. The beautiful thing about scheduling software is all of these sorting options are available.

Do not overlook the power of sorting in scheduling. When I did schedules, the sorting feature allowed me to understand the flow of work and the strategy for project completion. For example, I might sort the project by resources to ensure that every line item included a resource, or I may evaluate all the line items that require labor and review the calculations. Another thing I like to look at is definable features of work to assist in writing quality control/safety plans. The options available through sorting are endless and valuable.

Congratulations! The project strategy, time, the money, every step needed to complete the project is all wrapped up in this one document. We have just a few more steps before this proposal goes out for an award!

CHAPTER 5

Identify Resources

The bidding process for construction is ridiculous. If you do not believe me, call over five different contractors and request a bid to renovate your kitchen. Make sure you provide all of them with a wage schedule, the same plans and specifications, and even a specific material list outlining the exact products you want in the renovated kitchen.

All five of your contractors will submit bids and most likely, the bid for performing the work will vary significantly between the contractors. Based on the information provided, the variances should not be material, I mean they pay the same labor, buy the same material, and the work is pretty straightforward. Nonetheless, the range of numbers makes you think they bid different projects.

Unfortunately, contractors and subcontractors do not necessarily bid with accuracy as the primary objective. The bid incorporates variables that have nothing to do with the project specifications, but it gets thrown in with the number. Below are some examples:

- Too much trouble to review, throw a number at it – no review.
- Massive contingencies because they are too lazy to review the documents
- High bid – not really interested in winning
- Mark up on top of mark up
- Bidding based on expected bids from other subcontractors

Before I review these variables in detail, remember your job is to obtain accurate numbers for the bid. You cannot trust others without verifying what they provide, or you will lose a lot of projects due to your supply chain's faulty bidding. Verify!

In my previous life as a contractor, I would call potential subcontractors with a project and request a bid. Sometimes, within minutes of my conversation, they would send over a number. When that happened, I instantly knew they did not review the 2,000 pages worth of documentation I forwarded. There is no way they understood the bid, nor did the number they provided include the level of attention I required to win the project. When you ask a subcontractor/contractor for a bid, explain to them you want a breakdown of labor, material, and equipment. If they are not willing to provide this level of information, you might need to look elsewhere. Remember, this is your bid; you must understand all of the numbers.

On a project in Northern California, my previous company sent out a request for bids on a half-million-dollar job. Three separate electricians submitted bids, but the one I wanted to use came in significantly higher than the other two. As always, if you get one outlier, find out what they were thinking, which will confirm the others did not leave something out of their bid. When I got on the phone with him, he explained his number and then told me he added a twenty-five-thousand-dollar contingency because he never worked with us before. I met up with the guy, provided him with our bonding information, and he dropped the contingency from the bid. Again, research the numbers, understand the pricing, because your success depends on it!

By far, my biggest pet peeve, submitting a high bid to avoid winning. Sadly, this happens quite frequently. If the numbers come in high, make sure you evaluate the market carefully. Sometimes, the numbers received for the bid are inaccurate because the subcontractor is not interested in the project. Sometimes, depending on the market, all of the proposals may come in inflated. If this happens, then proceed with what I call the Wal-Mart - Kmart strategy.

When Wal-Mart wants to buy a product, they call you and say, here is what we will pay. On the other hand, Kmart calls the vendor and says, what will you charge. Simply put, the inflated number at bid time will reduce when the project is awarded. In the bid process, you might need to make some adjustments on proposals based on the assumption that prices will go down at the time of the award. To make these types of decisions, make sure you understand the market and demand.

Mark up on top of markup is when a subcontractor adds a margin to each line item and then adds a margin on the total of all those line items. For example, I reviewed a bid once on a Davis-Bacon wage project, and I noted that all of the labor lines were twenty-five percent higher after adding a burden. When I spoke with the subcontractor, he told me that he only charged a twenty-five percent margin for overhead and profit. Of course, he failed to mention he charged it on each line item, then totaled all the line items and marked it up again. I would love to tell you this is most commonly a mistake, but sometimes it is a deceptive process to increase margins. Again, the objective here is to identify actual costs without any fluff so that you can win.

The expectation is simple. The owner/general contractors send over the documents, the subcontractor reviews and respond with a price to complete the work. Unfortunately, that is not what happens. Subcontractors try and figure out what others are charging to do the job, and instead of providing an accurate price, you end up with a guess based on what others might bid. Look for this in the bidding. Make sure you evaluate every bid submitted carefully as if the award depends on it - because it does!

Earlier we discussed the importance of strategy and reviewed general strategy introduced by the King of Strategy, Mr. Michael Porter. Mr. Porter provided strategies, but he also went a step further and introduced what he refers to as the "five forces" that impact industry. One of the forces he discusses at length is the "bargaining power of the suppliers." His point is simple; if suppliers get too powerful, they ultimately control the destiny of your project, including the margins you hope to obtain. His position is well discussed as it refers to manufacturing, but I am going to take it one step further. If you let suppliers run your project or control your bidding process, you have succumbed to this power, and the project becomes difficult.

One of my core values is, "Never let anyone run your project." The reason this core value exists is that I observed so many projects where a supplier, client, or other third party took the reigns and began to control the direction of the project. This frequently occurs in the bidding process as team members get hooked with only one subcontractor, and then, at the last minute, they drop out of the running or submit high numbers and will not return your call to discuss. When a supplier responds in this manner, they are hijacking your project. The next question then becomes, "How do I avoid this happening?" I am glad you asked.

First, never trust anyone in this industry. Subcontractors sometimes have the best intentions, but things come up, and the work they currently are doing might take precedence over your project. Never depend on just one subcontractor or supplier. If you are pursuing a job in an area with only one subcontractor in a critical trade, you might need to evaluate this during the go/no go phase. Always find other bidders and remember you must manage the process, which means you have to check in on bidders periodially to ensure they are putting still bidding and working on your proposal. Don't just assume because some person you don't know said they would bid something that they will actually pull through.

Second, never accept their numbers at face value. Ask lots of questions to help you understand how they came up with their bid amount. It always surprises me how team members accept a proposal from a subcontractor and take it at face value. Remember that most subcontractors do not bid as accurately as you would like them to; for the most part, they don't have the time or resources. Also, don't forget that they can still bid after award to whomever won the project. Make them break down the labor, materials, and equipment so you can compare it to your schedule, research, and other subcontractor numbers. Also, ask who else they are bidding to. This enables you to understand your competition and if you are getting the best number. For example, if they tell you they are bidding to another contractor they work with frequently, you can assume your number is a little higher based on their relationship.

Third, make sure you pursue the critical subcontractors vehemently. Recently, I worked with a team that bid on installing a metal building for the government. The significant part of the project consisted of buying a pre-manufactured building, but, ironically, guess what area the team did not pursue diligently? That's right...the last day they were frantically searching for pre-manufactured builders.

Fourth, tell them exactly what you want from them. Don't call the electrician and say, "Here are the plans, specifications, and other documents." Do your due diligence and point them in the right direction. The more detailed you are, the more likely they will respond in kind.

Fifth, make sure you read their proposal. I am amazed at some of the exclusions included in proposals. I remember reading one that said they required 100 percent down upon acceptance of their purchase order. I have read others that excluded things that are required by law, such as wage requirements or safety standards. Read everything they submit, and if they have exclusions, note them and document who has that covered or if it is something they need to change and resubmit.

Finally, and by far the most important, do not hire subcontractors who subcontract out their work. When you hire someone who hires someone else, you are giving away valuable margin and most likely will come in high on the bid. Make sure you do your due diligence on the subcontractor and make sure they have crews and self-perform the work you are asking to be done. The last thing you want is a middle man.

The most exceptional project strategy is quickly eliminated if the wrong subcontractor ends up on your project. From the beginning, it must be clear that this is your project and you make all of the decisions. Yes, it is a team, but you are the captain, and everyone must understand that implicitly.

In my thirties, I experienced an epiphany leading me to open a tree business. Stop laughing! I thoroughly enjoyed the work of pruning trees, clearing land, and doing minor landscape work. I even tested and received my arborist certification, which, by the way, is one of the hardest tests I have ever taken.

In the beginning, I spent most of my time on the project making sure we did everything correctly because the majority of projects came from word of mouth. A client referred us to a senior living community, and it gave us several months of work. Every day a neighbor would walk by and ask if we could complete some tree work at their house. One particular lady needed a lot of tree work, and because I was on-site, I got to know her pretty well.

Just about every day she asked if I would do some other work for her, including adding a Florida room to the back of her home. I kept declining, and she kept making the deal sweeter. Finally, I conceded and accepted the project.

Being naive (nice way of saying, "I am stupid"), I started my search for contractors to complete the work. I asked no questions, assuming subcontractors are professionals and would act accordingly. I reviewed the numbers and selected a crew based strictly on pricing.

The project started great. My team prepared the site, brought in the dirt, and prepared the pad just like my concrete guy instructed. As planned, I contacted him to let him know he could come and do his work. He never answered his phone and did not return my call for three days. Apparently, he decided to go party with the fifty percent I gave him as a deposit.

As you can imagine, I struggled with this guy for the rest of my project, ruining my relationship with the elderly lady and losing my reputation in the community. One wrong move with a subcontractor, and you lose your project! Never give up leverage, the money, until the work is complete. Never, ever pre-pay for anything in the construction business.

Vetting – Vetting is a term most of us think of when referring to a candidate for a political office. I remember this topic became quite public many years ago when John McCain selected Sarah Palin as a running mate. They quickly pushed her through and missed a lot of facts that ended up hurting his campaign. Choosing the wrong partner can lead to devastating results like the loss of an election.

Vetting a subcontractor is no longer difficult. The internet provides unlimited information, and they typically have a website that highlights clients. Also, you can request they provide you with some past performance, and you can go online and examine their state licensing to find out if they have any complaints filed against them. These simple procedures might save a lot of problems down the road.

Make sure you do not ignore the blatant signs. For example, if the bid submitted is in an email, not on a letterhead of a separate document; they use a personal email instead of a business URL; they have no office and are nonresponsive; and they regularly ask about down payments. These little clues might give you enough information about a subcontractor to avoid them despite the great price.

Finances – Rule number one for contracting: never pay for anything unless you receive something in return. When you make this statement, everyone nods their head as if this is just practical sense. Unfortunately, it happens more than necessary, putting a lot of contractors in a less than desirable position. This also means making sure on every pay application you only pay the portion of work that is entirely driven by—you guessed it—the schedule. When you pay out, it is in direct accordance with the percentage complete on the schedule; therefore, if the line item is 50 percent done, you distribute 50 percent of activity amount.

There are a few scenarios when pre-paying might be necessary, but it must be handled correctly. First, never forget you are a contractor, not a bank. Your primary service is not financing for small businesses or floating material for subcontractors. If a subcontractor tells you they cannot afford to pay for material, you renegotiate their subcontract agreement, removing the profit and overhead they added on the material, and buy it directly from the supplier.

As a small business, it is inevitable to deal with subcontractors that experience financial hardships. Unfortunately, this happens mid-project, so there are limited options. I worked with one contractor on a project that extended way beyond the original time frame, and many of the subcontractors on the job needed assistance. The contractor worked with each vendor paying for materials and assisting them with meeting cashflow to remain viable and finish the project. I got involved about mid-stream of him doing this, and when we discussed it, I said to him, "These subcontractors must love you because you are paying their direct cost (funding their project), and they pay no interest, and you allow them to keep the overhead and profit they originally charged you in their bids." If you must help a subcontractor, you must receive compensation for doing this just like a bank would charge for loaning them money.

Another situation that occurs quite frequently is the payroll dilemma. The subcontractor calls you and says, "I want to finish my work for you, but I ran out of cash and cannot cover my weekly payroll." He proceeds to tell you that if you can just pay him enough to cover his weekly payroll, he will keep working and get this job done for you. Unfortunately, contractors agree to these terms and pay him weekly off of an invoice. Never do that!!

If this occurs, you just set up an arrangement where each week he prepares payroll and submits the reports directly to you for review. You review these reports and confirm in writing with his employees that this information is correct. You then calculate the exact dollars he needs to cover payroll, determine the interest on this loan, and prepare an amendment to his subcontract agreement reflecting the charges. He will then sign this, and you can release the exact amount necessary to cover his payroll. Make sure immediately following the release of funds you get signed affidavits from his employees that they were indeed paid.

Make sure you document everything! Don't release payment until every release is in your possession for all the parties. Make sure you clearly explain all of this at bid time and in your subcontract agreement. No exceptions!

CHAPTER 6

Evaluate Competition

Let me begin this chapter with some perspective on competition. First, when evaluating competitors, always look at it as a motivator, not a measuring process. Do not look at the competitor and judge your operation against what they do, instead view their methods and let it act as a motivational tool for you to improve. A core value I preach all the time is competition is internal. Simply put, we are our own worst enemies. If you want to resolve an issue within your company, look internally, that is how you address challenges and make your company better.

In bidding, it is imperative to understand what other companies are pursuing the project. If you know the players, you can anticipate the strategy they intend to use to win and prepare your game accordingly. As discussed earlier, the key to winning projects is a unique strategy. If everyone pursues the same plan of attack, then it is going to be a crapshoot.

Who and how – Do you remember the various team positions we discussed earlier? The wild card is the person that shows up on bid day and asks all kinds of random questions to make sure the team did not miss anything in the bid process. As a wildcard, I always ask how many other people are bidding on the project? Do we know them?

When I first started bidding with my new team, the response I received to the question is the deer in headlights look. As they stared, I would say, you mean to tell me for the last two weeks you have talked to tons of suppliers/vendors, and subcontractors and no one thought to ask who else might bid the job? Then the nod comes. The bid manager should be asking these type questions daily to keep everyone in the investigative mode. Honestly, on just about every project I ever bid, if you ask questions, you will find everything you need to know about the other bidders. We are not searching for proprietary secrets, just general facts. I like to know who I am battling!

Here are some sample questions:

How many other contractors are you bidding too?
Who else is bidding this job, do you know?
I noticed you were on the job walk, do you intend on bidding the project as a prime or subcontractor?
Have you ever worked with any of the prime's bidding this project?
Are you a prime contractor?
Do you intend on bidding the job to multiple companies?

Simple questions providing a lot of information.

What to do with the information – Data is priceless. Once you start understanding a little bit about the competitors, then you can evaluate what you are doing. For example, consider bidding a job in a remote location several hours from the nearest city. Because of the remote location, the assumption is all bidders will need to add per diem to the cost. While calling about supplies, you find out through some simple questions about a local contractor. Guess what? You need to know a little about his operation to decide if you can overcome the apparent advantage he has as a local. Maybe you can overcome the overhead by a small adjustment in the margin. The point is understanding what you are going up against assists in vetting the strategy.

Ultimately, with a little research, you might figure out the strategy of the other contractor and figure out a way to push them out of the picture by doing what you do best. The objective is not to sit back and worry but instead to understand your competitive advantages versus his and leverage them.

Let me give you an excellent example to make my point. During the recession, many large contractors began pursuing projects not generally in their portfolio. We pursued a project at the airport and sent one of our project managers to the job walk. When he returned from the visit, he looked discouraged. He said, "Well, we are done...such and such contractor is bidding on this project." I started laughing because I could not believe he had already conceded without even a fight.

I took advantage of this moment to teach my team about understanding competition and strategy. I explained that we had lots of internal benefits not available to them strictly based on their size and structure. Since we knew they were bidding, we can structure our strategy to leverage our strengths against their weaknesses. The more we talked about it, the more the team got excited about the opportunity to take down Goliath.

Periodically, the government releases a project that just sounds fun. To this day, I love heavy equipment. I am mesmerized by it and enjoy watching a good operator on a job. In 2011, the Air Force released a demolition project requiring destruction of about 40,000 square feet of buildings.

We followed the process in this book and began working on winning. After some due diligence, we identified our competitors: two large contractors and one individual that self-performed the work. The two large contractors would subcontract the workout, so I did not even think about them, but the small guy could easily win. I also knew the small guy did not really care because if he did not win as the prime, he would just give his number to whoever won.

We had to win this project! Our original strategy had us bidding it to self-perform and then subcontracting a portion of it out to a demolition contractor that we knew. Quite frankly, knowing the competition, I knew this would not work. We had to figure out a way to compete directly with this small guy, or we were out.

I decided to contact the small guy and see if he would bid to us directly and drop out of the running or at minimum give us a better number than the other two large contractors. He would not drop out, nor would he go exclusive, but he did agree to bid to us. I then started asking questions, "How long will it take to complete the work? How do you intend on completing the work?"

He laughed and told me they had done this for twenty years and were quite confident in both their time frame and methodology. It would take approximately four months to completion, and he intended on tearing it down and filling dump trucks to haul off the debris. No magic sauce on this deal.

I began drilling my team for ideas. The only way to win this project was to devise a manner of accelerating the schedule. (Remember the whole time and money bit; if you want to win consistently, always keep these two concepts at the front of your mind.) I took the prepared schedule and began studying it, line by line, attempting to determine what pushed the time to four months.

About 10:30 at night it hit me like a ton of bricks (demolition joke)! The time was in hauling off the material. Five or six dump trucks a day running continuously are not capable of removing the debris within the same time frame as demolition.

The next morning, the team and I knew what we needed to do to win. We needed to figure out a better way to remove debris more quickly, thereby reducing the overall project time. Finally, it hit us. Why not surround the site with dumpsters and fill them as demolition is completed. The trucks can keep emptying the dumpsters while the operator keeps on working. We set it up on the schedule, and sure enough, this knocked an entire month off the completion date. Bottom line, we bid this to finish in three months instead of four, and we won! We beat the other contractors by right at $100,000, and we ended the job with an 18 percent margin.

CHAPTER 7

MEETINGS

Most bids from start to finish take about two solid weeks to prepare. From my experience, even when more time is available, it still takes approximately ten days. One point to remember, do not forget about the cost to prepare a bid. If it cost $15k to win a job with a margin of only $5k, then it might not be the best job to pursue. One question I asked my estimating team when considering a job, how much will it cost to bid the job?

Oops, got distracted. Back to meetings! No one likes meetings, but everyone complains about communication. It never ceases to amaze me how much information never gets shared in the bid process. Let me tell you how it happens. After you win a job and there is some trouble, inevitably somebody says, "I knew that would happen, I was going to say something..." I am not going to lecture you on meetings, but make sure to respect people's time, provide an agenda, and allow everyone to speak. When I end a meeting to this day, I look at every person in the face and ask them to provide closing comments. I want to know what people are thinking, and so should you!

In the course of preparing a bid, communication must remain a priority. I provided a list of meetings conducted within our bid process. Depending on your work environment, you may need to add more or deduct from the list. In my office, we all worked together and communicated all day long. Sometimes we did not have formal meetings, but we talked all the time. Mold the meetings to your environment.

General overview meeting – Right after the bid team is selected, the bid manager should host the first meeting. The first meeting is where each person on the team introduces themselves and gives a little insight into what they already know about the project. The bid manager must make sure everyone engages during the first meeting. Do not allow anyone in the meeting to control another individual, or speak negatively about the project, the company, or the other team members.

In the meeting, it is a good idea to explain the process and define the players' roles. Make sure the team performed the necessary due diligence and understands the project. If not, tell them to gear up, read the specifications, and follow the guide provided earlier in the book. Most importantly, create some energy and excitement with the team.

Brainstorm – I think this is one of the most lacking traits in all companies. If you do not believe me, get up right now, gather all your employees in a room, and ask a question about the business, and watch what happens. Not much, very quiet. Dismiss everyone, and then they will all talk to one another and share everything!

We learn from an early age to guard against embarrassment from speaking out. Then we head off to college and the same deal, do not say anything for fear of shame or ridicule. We get our first job, and whenever we talk or have an idea, the boss ignores us or makes us feel stupid, or never asks what we think. BREAK THAT MOLD! Give your people the freedom of expression. Encourage it, push even those that will not share, make them get out of the box. Do not let other people ridicule or shut down the opinions of employees, help them speak out, which in turn provides you with valuable information.

Strategies come directly from creativity, and if you want a creative culture, then you must breed creativity. Get the team talking immediately about the strategy for the project, write down everything that is said, then come back the next day and let the team evaluate the ideas and figure out what will win the job. Conduct these type meetings as frequently as necessary. I ended up doing these type meetings at least twice a week with my team.

WBS – The WBS establishes the outline for the project. Make sure everyone participates in the meeting and contributes to building the framework. If you do not get buy-in from the entire team, it is going to cause difficulties later. Everyone should understand the workflow after the meeting. If any team member does not follow the outline, find out what is missing. If you need to, take a break and revisit the WBS, but when you return, make sure the team is on the same page. If the team cannot agree on the WBS, it is time to take a hard look at the project to ensure it is a good fit for your company.

Strategy – The strategy meeting is the final review of both the plan and the developed WBS. The strategy meeting takes place after preparing the WBS. Once the outline is available, the team can now formulate how to proceed with delivering the project. The session should be fun, with conversations going back and forth between team members. If there is no dialogue, the team probably did not put a lot of thought into the project. The last thing you want to do is develop a strategy that you must change on bid day. Take your time on this point and get it right.

Assignment Meeting – Assigning duties is something that needs a daily review. The Who? What? When? report is the bid manager's best friend, and the entire team should fully understand their responsibilities and the required timeline. Clear direction is essential in the management of any process; therefore, the bid manager must maintain this report and review it continually with the team. Normally, immediately following the strategy meeting, the assignment meeting occurs, and the team naturally assumes different jobs. The bid manager simply documents what everyone agrees on doing and then holds them accountable.

Project Scheduler – Normally, the project manager assumes the responsibility for preparing the schedule. Sometimes, project managers might not possess the necessary computer skills to enter the information, but they do know how to write down the flow of work. If this is the case, simply ask the project manager to use a piece of paper and document the process. Then enter the data into your software and have him approve it as entered. Do not get hung up on software issues; remember that the objective of software is primarily to assist the user.

It is imperative that the bid manager track the schedule preparation. This is the key document for the bid and must be prepared correctly and in a timely fashion. Many projects go south quickly because the project manager does not complete the schedule in time to support the rest of the team. As stated before, this is the "bible" of the project. If there is no schedule, the entire bid system falls apart leaving it up to fate!

As a side note, if the schedule is not completed on time, remember this little saying: "There are no bad Indians, only bad chiefs."

Daily briefing – The team should meet for about fifteen minutes daily. Each team member should update the bid manager on their duties and advise the team of any pertinent information. The bid manager should provide the Who? What? When? report and update it in front of everyone. The status of the responsibilities provides the bid manager with the direction they need to "manage."

The wild card team member attends these meetings periodically, asking questions and keeping the team thinking. The bid manager invites this person whenever they think the team might be losing focus or needs a little jump start.

Before each meeting, the bid manager should check the source for the bid to ensure no amendments or other new information is available. Do not assume someone else is checking this portal.

Subcontractor – Despite all my years in this business, I can honestly tell you I do not understand why subcontractors and prime contractors do not work together on bidding. I mean, think about it. Wouldn't you want to involve everyone in developing a price that includes major risk that might cost a lot of money? The process now is broken in my opinion. Prime contractors send over documentation, wait until the last minute, and then use the number not knowing if the subcontractor bid it correctly.

Here is a novel idea. Engage your subcontractors and conduct meetings with them to review the scope of work. If local, invite them to your office and involve them in the planning. If the strategy requires one of them to complete a large portion of the work, hook up with them quickly.

Here is a fact: most of the time a project is lost not because of the prime's numbers (profit, overhead) but rather because of inaccurate pricing provided by the subcontractors. Make sure you include them as much as possible in the process.

Pre-Bid day test – Normally, a day or two before the bid due date, the team needs to perform a pre-bid test. This test includes running through everything to identify holes or gaps in the numbers or other pertinent information. It provides the team an opportunity to question one another and ask other team members questions related to their assignments. Also, the bid manager needs to be reviewing all the assignments to ensure completion. For example, there is a certain task that is just not covered, so the bid manager asks two team members address it and create a viable solution. This meeting is intense because the smoother this meeting goes, the more likely you are to have a smooth bid day process.

Bid managers, this not the time to accept excuses or grant leniency. Your team must understand the importance of this day and the related assignments. About a day before the meeting, review in detail everything that must be done before this meeting and do not accept excuses. Anything not ready now flows into the final bid day and leaves you no time for processing or review.

Final Strategy Review – The final strategy review occurs immediately following the pre-bid test meeting. In this meeting, the bid manager reviews all aspects of the project to ensure alignment with the overall strategy. It amazes me how team members lose direction during the process; therefore, the bid manager must review all aspects to make sure there is continuity between the team members' work. For example, if a proposal narrative is required, a team member should review making sure it aligns with the estimate and the project schedule. The proposal should also reflect the strategic approach the team seeks to execute and communicate it clearly to the agency or company. At this point, it is about ensuring all documents point in one direction.

Bid Day – Bid day finally arrives! The bid is submitted either electronically or it is mailed out. Not so fast. This is by far the most important day because typically this is when you finally collect all the data, including the numbers from some remote suppliers and contractors. The most important thing about today is that the entire team needs to be available for review. Many times, one person on the team did not finish all their assignments and spends the entire day scrambling to pull their part of the bid together. If this happens, the bid manager did not do a good job of organizing and inspecting the status of work. Deal with that after the bid.

In my former company, we set up the conference room with a desk for the various team members, a television screen for viewing documents and each other's screens, and a telephone for phone calls. We referred to this room as the war room! Each of us evaluated the schedule, going line by line and discussing the risk, the pricing, and the strategic plan for each activity. Team members displayed calculations and discussed information related to each task, and the wild card engaged and challenged every premise presented. The wild card's job is extremely important on this day, as he or she attempts to rattle the team members with questions about their calculations, takeoffs, assumptions, etc.

Through a series of questions, the wild card begins to evaluate whether the team is confident in what they present. As a wild card, I regularly attempted to find areas in the proposal that included assumptions instead of accurate analysis and bring those out in the light. I also questioned strategies and attempted to argue against their effectiveness on this project. The objective is to see if you can punch holes in the work of the team. The better the strategy, the less rattled the team gets, meaning they did the homework and the approach is solid.

Make sure every team member engages in this process. The meeting is normally pretty intense as each person on the team defends their work from the last two weeks. Despite all the controls, normally the wild card identifies serious mistakes in the thought process, calculations, and assumptions.

Bid day arrived, and my team had worked diligently over the last two weeks to prepare for a project located in Northern Nevada. If awarded, it would be the largest project this team had ever acquired. The project was set up as an IFB (invitation for bid) and required no proposal, just one single number to determine the awardee. My involvement thus far had been limited to a few conversations and review of general information. I arrived at the office anticipating an easy review and a quick submission. Honestly, this was one of those times I questioned the idea of the wild card, thinking that position might be of no use with the right group of professionals.

The meeting began, and I listened intently to the developed strategy and observed as the various documents showed up on the big screen. I started asking some simple questions and observed some very odd mannerisms from the lead estimator. He answered my questions, but with hesitation. I started pushing the other team members, and I noticed similar reactions. I quickly realized that the team members had not coordinated on preparing the schedule, nor had the bid manager maintained a level of control on the activities necessary to complete this bid. Quite frankly, I thought this was not going to happen, and we would never get this submitted.

Fortunately, the project schedule did make sense, and it included the current pricing and resources. I pulled up the schedule and drilled line by line through the entire document, adjusting prices, dates, and resources off the team's feedback. We argued, screamed at one another, and even left the room a few times to calm down, but we made it through the whole schedule and lowered the price by $175,000. We won the project by a slim margin, and to date it remains one of my favorite projects.

CHAPTER 8

Determine Risk Rate

Contracting is unique compared to other businesses for many reasons, but one in particular is the determination of the risk rate. By risk rate, I mean the money added to the cost of delivering a project, or profit. The reason I consider this unique is because in the contracting business profit is usually an afterthought by the owners at the last second before the bid goes out. Don't get me wrong. Many businesses do carefully strategize and determine a rate, but in most cases, it is merely a percentage tacked on top that is calculated more by subjectivity than rationalization.

As a cost accountant, a lot of my studies revolved around pricing strategies. After determining the cost, the next question is the margin added to create profit for the entity. As you can imagine, this process is quite intense and considers multiple variables, including the cost, capacity, the market, the consumer, and the competition. As a contractor, you should invest the same amount of time into determining what you anticipate as a margin for the company and each individual project. This section deals with profit for a project, but it is necessary to also create a master plan for the profit of the entity.

Risk carries with it a lot of connotations. It is one of those buzzwords that seem to come up in a lot of conversations regarding construction, insurance, and any other time someone attempts to scare you! Business is risky, and profits represent the reward for assuming a risk.

When I turned eighteen, I experienced my first taste of risk in the form of purchasing vehicle insurance. With all my information in hand, I called the insurance company. They asked me about fifteen different questions, told me to hold for a minute, and then returned with the bad news. "Based on your age, gender, and driving record your premium is...." I about dropped the phone after hearing the price. The lady on the other end knew she had just shocked me, so lovingly she said, "Son, insurance is taking on the risk, so you can drive without worry. As a young person, the risk is high; therefore, the rates tend to be higher."

Lesson learned: high risk, high return, low risk, low return. As I got older, married, and settled down, my policy premium dropped significantly. This was mainly because the risk of my driving poorly reduced, and my record (experience) indicated that my probability of wrecking fell. The insurance return on me is lower because I am not a risky driver. Other drivers with a poor record (high risk) demand higher premiums because the insurance company must compensate for the probability of an accident.

The risk then determines the profit margin for a project. Simply put, what is the risk of doing a project? If there is a high-risk factor, the margin should reflect that, but if the project demonstrates minimal risk, the number should reflect that as well. Contractors get in trouble on this because instead of thinking strategically they are focused solely on maximizing profits.

Maximizing profits is awesome! However, if you never win a project, it becomes irrelevant. You must consider the risk of the project and assign a margin accordingly. You might even think, *"Well, this is great, but for the last ten years I have been making a killing just marking it up."* That's great, but as technology improves and consumers become more knowledgeable, this surely is going to change. Those who understand profit/risk stand to remain viable while others slowly shed market share.

Let's talk through developing a solid profit margin. First, let me start by saying this book is mainly for small businesses; therefore, the structures I am speaking of are meant to assist the small company with limited resources. Large operations possess the resources to incorporate more intense pricing strategies and perform a wide range of analysis.

First, before you ever even approach a project, you need to understand your business margins. How much money do you need to break even? What is your anticipated overhead for a year? How much profit do you need to generate before taxes? These seem like reasonable questions, but I guarantee many reading this book possess no clue to these answers. Never fear. After reading this chapter, you will understand.

Creating the profit begins with answering the above questions. For starters, you need to understand what your annual fixed cost equals. To make it simple, these are the costs that do not change relative to your sales. For example, if you rent office space for $1,000 a month, this is a fixed expense and remains the same whether you win one or ten projects. Below are few examples of fixed costs.

Office Rent	$12,000.00
Administrator	$25,000.00
Insurance	$10,000.00
Utilities	$3,500.00
Payroll	$42,000.00
Total	**$92,500.00**

The next step is to attempt to project sales for the following year. Start by calculating how much carryover work you have from the current period. For our illustration, we are going to pretend there is no work on the books for the next year. With that in mind, let's discuss how we can project sales.

First and foremost, you must consider capacity. How much can you physically complete within one period? Look at your current personnel, the time available for bidding, past sales history, and the market. If your team consists of you and one other person, your capacity revolves around what the two of you can accomplish. If additional resources are available, then this may increase capacity. Evaluate your capacity against financial measures as well, such as bonding, cash flow, and debt load.

Second, perform some research on your potential clients. If you work for the same clients year after year, find out what projects might go out for bid. Also, create a marketing plan indicating whom you intend to work for and then try to figure out what they have planned for the next year. Most of the time, construction projects for commercial or government entities are planned long before the contractor sees them.

Third, study the market. What is the expectation of larger contractors? What does the financial market look like? What about the direction of the government? These types of questions are not always easy to answer; however, when you start doing some research, it is amazing the information available that can assist you in making projections. Do research on margins for typical construction projects and use this information as a starting point for your business. Most likely, the exact margins of competitors are not available; however, if you bid throughout the year and lost, you can reference this information to determine margins in your range of business. Consider time, bidding a job that spans over the years may need additional consideration.

Fourth, calculate how many projects you can pursue, the average price for those projects, and your current kill rate. For example, if you pursued ten separate projects with each project at a value of half a million with a 60 percent kill rate, your sales should be $3 million. Then, with this number, you work backward through the other suggestions, ensuring you remain within your capacity, client's expectations, and the market.

Finally, do the math to ensure the calculations make sense, and the numbers overcome the fixed expenses. What is the point of chasing a sales projection that does not meet the basic needs of the company? Unless you are a startup, your sales projection should at a minimum cover the fixed expenses. This final step is where you determine your margin based on the information collected.

Let's walk through an example to assist in understanding the concept. Following are the fixed expenses calculated earlier:

Step 1: Determine fixed expenses.

Office Rent	$12,000.00
Administrator	$25,000.00
Insurance	$10,000.00
Utilities	$3,500.00
Payroll	$42,000.00
Total	**$92,500.00**

Step 2: Perform research on clients.

After researching the clients, we work with, it seems they intend to release a lot of renovation-type projects this year in the two hundred thousand or less price range. The pricing aligns perfectly with the firm's sales objectives.

Step 3: Study the market.

When I say study the market, all I mean is do a quick review of what all the big companies are planning to do. It is not perfect, but you can lean on all their research to make your decision. For example, through research, you might realize the current administration is big on military spending, meaning there is work to be done!

Step 4: Calculate projects, amount, and kill rate.

Based on our kill rate from last year of sixty percent and bonding capacity, we can pursue six projects in the two-hundred thousand range. The company should do approximately seven-hundred and twenty thousand for the year.

Step 5: Perform calculations to ensure financial acceptability.

Now calculate the assumptions and make sure the project falls within an acceptable financial range.

Model 1:

Sales	$720,000.00	100%
Cost of Sales	$648,000.00	90%
Gross Margin	$72,000.00	10%
Fixed Expenses	$92,500.00	13%
Operating Income	–$20,500.00	–3%

As you can see, in the scenario above, the company loses money. There are a few options for consideration: increase sales, increase gross margin, and/or reduce fixed expenses. Most likely, growing sales or adding more work is the appropriate answer.

Model 2:

Sales	$720,000.00	100%
Cost of Sales	$626,400.00	87%
Gross Margin	$93,600.00	13%
Fixed Expenses	$92,500.00	13%
Operating Income	$1,100.00	0%

In model 2, the gross margin increased from ten to thirteen percent. The question becomes, do you remain competitive with the higher margin. By running various scenarios, the owner can see the options available for generating a profit. If the business has a bonding capacity, you can figure out how much-bonded work is needed versus non-bonded jobs.

Model 3:

Sales	$1,281,250.00	100%
Cost of Sales	$1,178,750.00	92%
Gross Margin	$102,500.00	8%
Fixed Expenses	$92,500.00	7%
Operating Income	$10,000.00	1%

In the final model, sales increased, the margin dropped, and the company nets approximately ten thousand. The owner knows the sales goal for the year and establishes an average gross margin for projects. The scenario analysis serves as a gut check for owners. For example, if the sales number to reach a profit is not realistic, you need to back up and figure out how to generate a profit.

In my current job as a forensic accountant, I run into contractors and other business owners all the time that have no idea what they need to do to make money. Seriously, they do not know what sales need to be, or if expenses are too high. What usually happens is they do nothing until they eventually run out of cash. Know your numbers!

These simple calculations assist management in making some key decisions for the future. There are other formulas for calculating break-even, but to keep things simple for a new business owner, this accomplishes the same task.

Once the company decides or projects the overall margin, then this can become project specific. Let's say the company's overall margin for the year is eight percent. Simply put, to cover fixed expenses and generate a profit, the minimum margin must be eight percent. When dealing with an individual project, the team knows the minimum margin. If the company already has a higher margin for the year, the team may have an option to decrease the margin if the project is one the owner wants to win.

One thing I want to remind you of before we get into the profit calculation, remember to markup line items based on the risk related to the task. I have won more projects by implementing this one simple trick. Just remember risk determines the margin, low risk, low margin, high risk, high margin.

Now that you know the minimum margin, you can establish what I refer to as the profit calculator. The day before the bid goes out, I have each team member complete the profit calculator to determine their perspective of the risk. The bid manager combines the percentages and establishes the baseline for the margin.

If a team member comes up with a higher risk rate, or lower than the rest of the team, determine what he/she is thinking. If the team collectively comes in lower than the minimum margin, figure out why and, if necessary, get the owner involved to see if going lower is acceptable.

Column 1 – The Factor

The factor indicates the item of consideration related to the project or the bid manager's top seven general risk categories. The next few paragraphs I will walk you through seven risk factors used on previous projects to determine the risk factor. These factors change based on what the bid manager considers most important as it relates to risk. Remember this ultimately determines the margin you intend to pursue for the project.

-Degree of overall risk – After two solid weeks of interviewing subcontractors, reading specifications, and creating a schedule and strategies, most of the team has a general understanding of the overall risk of the project. I would tell my team to think of it as an investment option. If I asked you to invest in this project based on your knowledge, would you do so?

-Relative difficulty of work – Difficulty of work is based solely on the company's past performance and experience. Let's say your company does nothing but build apartment complexes, and the project you are pursuing is for an apartment complex. Based on your experience, the work is relatively not that difficult. However, if this apartment complex is built over water, that might change the game. Again, based on the previous two weeks of work, the team easily rates this factor.

-Size of the job – Once again, this factor is relative. If the company is performing a larger scale project than normal, theoretically the risk increases. This depends solely on the size of projects the firm normally completes.

-Period of performance – The longer the project, the higher the risk. If the project exceeds a year time frame, this increases risk. Also, the team must consider the impact of a long-term project on internal resources. For example, if the company only has one self-performance crew and they commit to this project for the entire year, this obviously impacts the firm's ability to pursue other projects.

-Contractor's investment – This considers the time value of money on the project. If the contractor intends to fund the project through self-performance or through supplying a large amount of material, this is an important financial factor that deserves attention. If the contractor intends to fund the project, this risk factor increases significantly.

-Location – Where is the project? Is it located at a familiar site with accessible resources, or is it located remotely at a place the firm has never worked? Location may increase the risk of the firm.

COLUMN 1	COLUMN 2	COLUMN 3	COLUMN 4
Factor	**Rate**	**Weight**	**Value**
Degree of risk	26	0.05	1.30
Relative difficulty of work	10	0.03	0.30
Size of job	1	0.06	0.06
Period of performance	1	0.05	0.05
Contractors investment	5	0.03	0.15
Location	35	0.1	3.50
Subcontracting	22	0.1	2.20
	100		**7.56**

-Subcontracting – How much of the total job is subcontracted? Do you know the subcontractors, or have you worked with them previously? Did they provide bid bonds or any form of security to protect you in case they back out? Are there plenty of other options at the project location if a key subcontractor decides not to pursue the project?

Column 2 – Rate

Column 2 ranks each factor separately with the total equaling 100. This enables everyone on the team to allocate a percentage of risk to each factor based on his/her opinion. In the illustration provided, the individuals opinion is that 26 percent of risk on the project is associated with the "big picture" degree of risk, and 35 percent is due to the location of the project.

It is just a ranking of the identified risk factors from the most important to the least important or from the highest risk to the lowest risk.

As a bid manager, it is imperative you discuss the ranking with each person on the team. Many times, this exercise provides direct insight to an item that a team member does not feel confident with. For example, if all the team ranks the factor subcontracting very low, but one person has it high, this requires a little investigation. What is it making this one person so concerned about a factor while the rest of the team rates it as insignificant?

Do you remember some of your thoughts from your younger days? Hopefully, I am not the only one that recalls some of the less than intelligent thoughts from childhood. For example, I remember finding out a family friend worked as a manager. Through a school program, I joined him for a half-day to observe the job. At the end of the day, I thought what a waste he just walks around and talks to people, "Business is easy; just find people that do what they are supposed to do, and then you do not have to pay a manager." The minute I started working as a teenager, I realized the importance of management.

My team pursued a relatively large project in Northern California. The project included a two-volume proposal, past performance, and a very difficult technical proposal explaining how we intended on performing. The bid manager in place at the time assembled a crew of seasoned professionals who, in theory, required minimal oversight. During the two weeks leading up to the project, the manager ignored many steps in the process trusting that the professionals on the team individually understood their respective responsibilities. She asked questions of the team, but she never inspected the work or pushed on deadlines, assuming all the time that each person had it handled.

Two days before the bid, she reviewed the project schedule, the proposal, and the resource schedule. Despite the noted lack of communication between team members during the process, the documents looked perfect. Based on her assessment of the team over the last two weeks, she dropped her role as bid manager and became one of the team members working on takeoffs and other project-type activities. Knowing she needed to at least appear to follow the process, she sent out the profit estimator and requested the team members to respond.

The next morning everyone on the team responded. She forwarded the results to me in preparation for the bid meeting. I opened all of them and noticed the project manager rated the period of performance at 50 percent. Everyone else on the team ranked this factor at below 5 percent. First, I assumed an error and decided to wait and find out when I got to the office.

I arrived, and after sharing the normal niceties I asked the obvious question. Much to my surprise, the project manager replied, "The entire schedule is a joke. It is not right. I have no idea if we can finish the project within the time frame, nor do I feel confident that it is even right." Ironically, as you are probably thinking, she was the one who built the schedule. The bid manager just stared into space; the rest of team became very quiet. My thoughts were that the bid manager did not manage the process. For the next twenty-four hours, the project manager and I created a working schedule. We did not win the project, but it did reinforce my position on management!

Column 3 – Weight

This category uses the margin to weight the factor. For example, if the company sets up a 10 percent margin for this year's projects, the weight follows the scale below:

.01 0.1

Each factor is rated and then weighted from this scale. If the risk for the factor is extremely high, it is a 0.1. If the risk is low, it is a .01. Of course, there are multiple options in this range to reflect the risk assigned to the factor.

Just think of it like grades from high school. A teacher gives you every assignment with a score ranging from 1 to 100 and then tells you that quizzes, tests, homework assignments, and projects carry a different weight. The grade is then a product of the rate times the weight.

Column 4 – Margin

The last column is the product of column 2 and column 3. At the bottom of this column you see the profit percentage for this project. If the margin is extremely low, that indicates a low risk project. If the number is closer to ten percent or greater, this indicates a higher risk job and pushes the margin up to the top-tier percentage.

Once the team submits their individual analysis, the bid manager reviews the input, prepares his or her final profit, and seeks approval from upper management. Once decided, this is the rate utilized to assess every activity on the entire project. The bid manager needs to thoroughly review these profit estimators and conduct discussion with the variances. If the numbers are not relatively close, you need to take the time to understand why someone thinks differently. This might provide you some very valuable insight.

The intent of this entire demonstration is not to force you into using my method for profit calculation but rather to create an awareness of the importance of profit as you determine your final price for the project. Every aspect of the bid process must be strategic in nature, addressing the variables faced as you prepare the absolute best number hopefully leading to award.

CHAPTER 9

BID DAY

Bid day arrives! Typically, if doing a government proposal, the bid day already changed five times, but the day is here! Bid day is for the bid manager, he/she compiles all of the documents, performs final reviews, discusses strategy, makes the final adjustments, and sends the bid out.

Yea, not so fast. By far, the bid day is the most stressful part of the entire process. If anything is going to go wrong, it happens on bid day. Here is the deal, listen carefully, the closer you stick to the process outlined in the book, the less stress experienced. Most of our bids had to go out before two in the afternoon, so let's go through bid day, assuming a two o'clock submit time.

First, the afternoon before bid day, speak with your team, and outline the expectations. If you do not conduct the conversation, inevitably someone will tell you they have an appointment or some other deal on the day of the bid. Plan on getting lunch brought in and tell everyone exactly what time they need to be at the office. No exceptions!

The morning of, start by addressing each team member advising them of their responsibilities. Of course, all of them will swear everything is done, not true. As a bid manager, it is imperative to monitor the team carefully to ensure no surprises. Do not believe them when they tell you all is good, ask questions, listen to the conversations, and jump in if necessary. The bid manager's focus is on the team, not task related to the bid.

At around 10 am, approximately four hours before launch, the team should assemble and begin a review of the numbers. With the schedule up on the big screen, the team should start reviewing every line item, discussing strategy, pricing, risk, and holes. No phone calls not related to the bid, no breaks, nothing unless specific to the bid - one-hundred percent focus on the bid.

Before going further, let me comment on a potential issue. The final four hours should not be the time team members or finishing up work that should have been done before bid day. If you observe that happening, review the bid manager's oversight.

Hopefully, in the "war room" there is either a television screen or some type of electronics allowing the entire team to view the project schedule. Post the schedule on the big screen and begin your review. The next several paragraphs go through the different filters on the schedule and how they are utilized to perform an extensive review of the project before the deadline to send it out. As discussed, this is something that should have been done in the pre-bid meeting and throughout the proposal/bid process. Today is different because it is game time!

The first filter is the WBS (work breakdown structure). I typically begin with this filter because this is the one everyone agreed on in the beginning. It enables the team to quickly go through the outline of the work just to make sure everything is covered. Ask the project manager to stand up and walk the team quickly through the various sections. While he does this, the bid manager runs a quick background check on the schedule to make sure there are no rogue line items that do not include codes or do not include a predecessor or successor. The team identifies and corrects mistakes and, if necessary, adds or removes line items under the various WBS categories. Once this review is done, the team is ready for the real fun.

Once the WBS review is complete, it is time for the wild card to step in and take over the rest of this meeting. She is going to quickly go through everything in the schedule, questioning the strategy, the schedule, and the entire bid/proposal. The objective is to really test every aspect of the project and identify any serious holes or weak spots in the strategic plan.

The wild card begins by doing the very same review just completed, applying a basic common-sense approach in her evaluation. She examines the various line items, observes the chart, and then just starts asking question about workflow, timing, and the amount of activities. She also asks the team to clearly define the strategy and then asks where this information is displayed in the project schedule. On one project, I served as the wild card and requested a breakdown of the strategy. The bid manager laid out a beautiful concept that made a lot of sense. I then asked her to show me how this played out in the schedule. Guess what look I got… that's right, deer in the headlights. Make sure the strategy is obvious in the project schedule. Remember, this is the crucial element that sets you apart from your competitors.

Next, the wild card filters the schedule by resources. The resource filter aligns the various activities with the actual individual or subcontractor who is in place to complete the activity. The wingman then focuses on those line items that involve the contractor, such as general conditions, supervision, quality control, and safety. Going step by step through these in-house items normally results in a reduction. For some reason, personnel tend to overestimate the required resources, increasing the cost. I remember one specific project where the estimator included $3,000 for fuel to get to the job and return after completion. Do you know how far you can go on $3,000 in fuel? Another common mistake is they do not allocate the superintendent's time correctly. For example, on smaller projects the team might set up the plan to get the superintendent to perform various tasks throughout the project. When they do this, they add money to the line item to pay the superintendent, but they also have a line item that pays the superintendent to be on-site throughout the entire project.

After going through the in-house line items, the wild card starts picking apart the subcontractor proposals. She reviews the line items and then requests to see a copy of the bid submission. Once in hand, the comparison begins. All she is doing is asking common-sense questions to determine whether the team has done due diligence. For example, on projects I will usually ask questions like, "How many days is this subcontractor on-site?" "How much labor is included in his bid?" "Did you ask if they are bidding to anyone else?" "How many bids did you get for this work, and do you have copies of the other bids?" Pick the schedule apart, drilling the team with questions about the subcontractors. The more questions you ask, the more areas you will identify that might salvage the project or get you a win.

One question that seems to always lead to a debate is about the subcontractor's breakdown of work. Estimators sometimes take numbers at face value simply based on past projects. When they do this, they rule out the analysis part of the estimate and take a risk that this estimate incorrectly reflects the work on this project. Every time, without exception, when I start requesting a breakdown, the estimator begins to complain and say subcontractors will not provide this information. Quite frankly, I disagree. Some do not, but if you ask simple questions, you can determine the breakdown and make sure it is logical. My favorite story of all time is a subcontractor who worked for us on several projects in Northern California. They did great work, and I really enjoyed working with them. After several projects, the estimating team in place began to just accept the bids without asking any questions. The contractor awarded a small renovation project to us that, per the schedule, took about four weeks to complete. We had this contractor coming in to do some work with the fire suppression system the last five days of the job. I looked at the schedule and noticed it indicated he came on-site after the small renovation work, did some testing, installed new sensors, and then completed a full test of the system. I asked the team to review his proposal and noticed the fee seemed excessive. I started probing for a breakdown, and after arguing for a bit with the estimator, he made the phone call. I heard him ask the question about how much money the guy had in labor, how many people would be on the job. Turns out, the guy had enough labor on the job to have three people on-site for the entire tenure of the project. If the estimator had not probed a little further, that guy would have made a lot of money for five days of work.

At this juncture, the wild card and most likely the bid manager understands the weaknesses needing resolution prior to submission. The bid is due in just a few short hours, so the bid manager assesses the required information and assigns it to the team members. Sometimes, this requires contacting subcontractors, re-evaluating internal takeoffs, or making a major strategy shift based on last-minute information or a lack thereof.

The wild card position is crucial. It is imperative you retain someone to complete this task who understands people and is good at putting them a little on edge. As the owner of the company, I served in this capacity for about three years. In my opinion, I had one of the best estimating teams ever; however, I can tell you on many jobs, after my interrogation, the bids dropped at times around 10 to 15 percent. The reality is these people spend two solid weeks working diligently on the bid, and because they are human, they sometime forget about the big picture.

The time has come! Get your most dependable soul to deliver the proposal bid to the post office with plenty of time for shipping. Let the wait begin!

CHAPTER 10

The Hand-Off (Transition Meeting)

The final chapter ends with transitioning the project from the bid team to the field. So, the question becomes, "I just won, what do I do next?" Glad you asked. After adopting the methods in the book, growth is on the horizon, so get ready!

After award, follow the line items in the schedule. For example, one of the first line items should be to obtain a bond; not only does it tell you what to do, but it also provides the person responsible. Of course, there are other administrative tasks due, which are outlined in the schedule. Once you get all the administrative work done, it is time to conduct the transition meeting.

The transition meeting is where the bid team hands off the project to the construction or field team. The idea is quite simple, get the bid team and the field personnel in the same room, put the schedule up on the big screen and walk through the whole project explaining the strategy, the resources, time frame for completion, and any other pertinent details. Open it up to questions and let the two groups discuss the project. If the field team has questions or does not agree with any aspect of the schedule or the strategy discuss it immediately. Do not let anyone leave the meeting if there are any open questions related to strategy or execution. Also, remember one of the bid members, the project manager, is going to be assigned to the project, so he/she should assist in the transition meeting.

Not only does the schedule provide the time, it also gives you a pre-made schedule of values. I want to go over this because it really connects the money to the schedule and customers/clients appreciate the congruence between the two documents. Every line item on the schedule with a cost represents a line item on the schedule of values. If I complete fifty percent on the line item A380, then I can bill for fifty percent. It makes the billing process simple, clear, and concise. Trust me, this is a great help for you as you struggle for payment throughout the duration of the project. Once the quality assurance person for the owner/obligee signs off on the schedule, the billing is simple, and no one can argue over the percentages.

The construction business reminds me of a three-legged stool. The three legs represent the three keys to a successful construction company: getting the work, doing the work, and keeping score. Each leg is equally important; if one leg fails, the whole stool falls over.

Understanding the three legs is why I created the system you just finished reading. My objective was to provide a tool that allowed contractors to efficiently manage the other two legs by executing the first leg. If you follow the process, the other two legs fall in line as the schedule assist with doing the work, and keeping score.

REFERENCES

Collins, J. (2001). *Good to Great* . New York, NY: Harper Collins.

Covey, S. R. (2004). *The Seven Habits of Highly Effective People* . New York, NY: Rosetta Books, LLC .

Fundera ledger. (2017, January 24). *funderaledger.* Retrieved August 2, 2017, from Small Business: https://www.fundera.com/blog/what-percentage-of-small-businesses-fail

Gerber, M. (2003). *The E-Myth Contractor.* Pymble, NSW, Australia: HarperCollins Publishers.

Harnish, V. (2002). *Mastering the Rockefeller Habits.* Ashburn, VA, USA: Gazelles, Inc .

Lencioni, P. (2012). *The Advantage.* San Francisco, CA, United States: Jossey-Bass.

Magretta, J. (2012). *Understanding Michael Porter.* Boston, MS, United States: Harvard Business Review.

Merchant, C. M.-C. (2017). *Management Accounting* . Montvale, NJ: Institute of Management Accountants.

Mubarak, S. (2010). *Construction Project Scheduling and Control* . Hoboken, NJ: John Wiley & Sons, Inc.

Porter, M. (1985). *Competitive Advantage.* New York, NY: THE FREE PRESS.

Printed in the United States
By Bookmasters